Rhetorics and
Technologies

**Studies in Rhetoric/Communication**
Thomas W. Benson, Series Editor

# Rhetorics and Technologies

New Directions in Writing and Communication

Edited by Stuart A. Selber

The University of South Carolina Press

© 2010 University of South Carolina

Cloth edition published by the University of South Carolina Press, 2010
Ebook edition published by the University of South Carolina Press, 2013
Paperback edition published in Columbia, South Carolina,
by the University of South Carolina Press, 2013

www.sc.edu/uscpress

22 21 20 19 18 17 16 15 14 13    10 9 8 7 6 5 4 3 2 1

Manufactured in the United States of America

The Library of Congress has cataloged the cloth edition as follows:

Rhetorics and technologies : new directions in writing and
  communication / edited by Stuart A. Selber.
    p. cm. — (Studies in rhetoric/communication)
  Includes bibliographical references and index.
  ISBN 978-1-57003-889-1 (cloth : alk. paper)
  1. English language—Rhetoric—Research. 2. English language—
Rhetoric—Computer-assisted instruction. 3. Communication and
technology. 4. Discourse analysis. I. Selber, Stuart A.
  PE1408.R54 2010
  808'.042028—dc22
                                                      2009039688

ISBN 978-1-61117-234-8 (ebook)
ISBN 978-1-61117-331-4 (pbk)

# Contents

List of Illustrations vii
Foreword
   *Carolyn R. Miller* ix
Acknowledgments xiii

Introduction
   *Stuart A. Selber* 1

1 Redrawing Borders and Boundaries

   Being Linked to the Matrix: Biology, Technology, and Writing
      *Marilyn M. Cooper* 15
   Among Texts
      *Johndan Johnson-Eilola* 33
   Serial Composition
      *Geoffrey Sirc* 56

2 Constructing Discourses and Communities

   Appeals to the Body in Eco-Rhetoric and Techno-Rhetoric
      *M. Jimmie Killingsworth* 77
   Unfitting Beauties of Transducing Bodies
      *Anne Frances Wysocki* 94
   The Rhetorics of Online Autism Advocacy: A Case for Rhetorical Listening
      *Paul Heilker and Jason King* 113
   Narrating the Future: Scenarios and the Cult of Specification
      *John M. Carroll* 134

3 Understanding Writing and Communication Practices

   Technology, Genre, and Gender: The Case of Power Structure Research
      *Susan Wells* 151

Rhetoric in (as) a Digital Economy
   *James E. Porter* 173
Literate Acts in Convergence Culture: *Lost* as Transmedia Narrative
   *Debra Journet* 198

Contributors 219
Index 223

# Illustrations

1.1. Source theories  17
1.2. Cognitive shortcuts  27
2.1. Image from "Text in a Machine" series  35
2.2. Strahav Monastery library  35
2.3. Ramelli's book-wheel  36
2.4. Petroglyphs in Chilas, Pakistan  39
2.5. Masses of mass-produced books  41
2.6. Work spaces of Johndan Johnson-Eilola, Dennis Jerz, and Charlie Lowe  42
2.7. Multiple views of one text in Tinderbox  43
2.8. Two views of Hitachi's µ-Chip  44
2.9. Partial listing of reader activity on a Weblog post  46
2.10. Ego surfing via a Technorati search  50
2.11. Scanning physical items into Delicious Library  51
2.12. Access of Web sites via eXtreme and Google Analytics  53
4.1. The earth-organism-machine continuum  78
4.2. Body divided from mind and bracketed with earth  78
7.1. Virginia Tech publicity collage  143
7.2. Web site of Spring Creek Watershed Community in 2003  145
8.1. Tools of the trade, 1960  154
8.2. Tools of the trade, 1976  155
8.3. Red River Women's Press, Austin 1973  157
8.4. Writing at the Freedom School, 1964  158
8.5. *Who Rules Columbia?*, cover  160
8.6. *How Harvard Rules Women*, cover  162
8.7. *How Harvard Rules Women*, table of contents  163
8.8. *How Harvard Rules Women*, first page  163
8.9. *Women and Their Bodies*, 1970, cover  165

8.10. *Women and Their Bodies*, 1970, table of contents 166
8.11. *Women and Their Bodies*, 1970, first page 166
9.1. The long tail of digital economics 177
9.2. Web 1.0 online version of Teacher Knowledge Standards 183
9.3. Web 2.0 social networking site for teacher-educators 185

# Foreword

Rhetoric, Technology, and the Pushmi-Pullyu

Carolyn R. Miller

In his recent book, *Saving Persuasion*, Bryan Garsten observes that there are two "forms of corruption" to which rhetoric is susceptible. These "twin dangers" stem from the very nature of persuasion, "which consists partly in ruling and partly in following" (2). In seeking to influence the beliefs, feelings, and attitudes of others, we may try too hard to rule, that is, to manipulate others for our own purposes. Or we may try too hard to gain their goodwill and assent, that is, to follow them by pandering to their presumptions and prejudices. In either case, truth and justice, cooperation and disclosure, will suffer.

This dilemma has the same structure as the push–pull model of technological development. In this model, technological change has two possible causes, supply and demand: it can be "pushed" along by the supply of discoveries and developments internal to technology itself (or derived from science), or it can be "pulled" by external forces, primarily market demand. Given the widely accepted premise that technological change promotes economic growth, one question that exercises economists and policy analysts is which of the two causes of change is more important and which factors government policy should target in order to promote economic growth, those that influence supply or those that influence demand.

Technology, like rhetoric, can both push and pull at us. Not only do "artifacts have politics," as Langdon Winner has claimed, they also have rhetorics. Technology pushes or manipulates us by requiring us to do certain things and in certain ways; our communication technologies, highlighted in this collection, push us to send SMS messages with no more than 160 characters or to access a point on a scroll or a magnetic tape linearly, in one direction; a library card catalog (remember those?) requires us to seek information using one search strategy at a time. A technology pulls from us, or panders to us, by reconfirming and strengthening our inclinations and propensities; blogging, for example, hooked into the already pervasive celebrity culture of exhibitionism and voyeurism, enrolling many eager participants in an activity that they welcomed without knowing in advance that they wanted it. Many mobile devices similarly gain assent and consumer dollars by offering information that we had no idea we wanted but that we then find it hard to live without.

The ways that technology pushes and pulls at us are called "affordances." In psychologist James Gibson's original formulation, affordances are what an environment offers to an animal, "what it *provides* or *furnishes*, either for good or ill" (127, emphasis original). For example, a given natural environment affords materials and locations for birds to build certain kinds of nests but not others. In the context of communication technologies, affordances take the form not of material properties or ecological niches but rather properties of information and interaction that can be put to particular cognitive and communicative uses. Thus a technological affordance, or a suite of affordances, is *directional*, it *appeals* to us, by making some forms of communicative interaction possible or easy and others difficult or impossible, by leading us to engage in or to attempt certain kinds of rhetorical actions rather than others. Affordances both enable and constrain, they both pull on us and push at us.

This pushmi-pullyu dynamic is central to rhetoric, and Garsten is only the most recent scholar to have highlighted it. For example, Donald Bryant's influential explanation of "how rhetoric works" has the same structure: rhetoric, he says, has the function of adjusting "ideas to people and . . . people to ideas" (413). The well-known exchange between Lloyd Bitzer and Richard Vatz regarding the "forces" in a rhetorical situation reveals the same dynamic. For Bitzer (the "market-pull" version), an objectively existing situation "demands" a "response"; for Vatz (the "supply-push" version), the rhetor creates the situation for the audience.

Rhetoric and technology share this dynamic and its twin dangers because they are both arts of design: they are both in the business of balancing innovation with tradition, of initiating change and then compensating for it. If rhetoric is the art that adjusts ideas to people and people to ideas, we might characterize technology as the art that accommodates the material world to people and people to the material world. This shared dynamic gives rhetoric and technology a shared ambivalence toward both tradition and innovation (or what Kenneth Burke called "permanence and change"). In an essay called "The Fear of Innovation," Donald Schön discusses this ambivalence in the corporate setting: "Companies want new technology, new ideas. We all know this. Then why do they fight so hard to prevent anything new from ever happening?" (70). He observes that "technological innovation disrupts the stable state of corporate society" (78), pointing to the changes in production, marketing, quality control, management, labor, and accounting that may be required by the introduction of a new product, or a new process or raw material. Resistance to innovation by labor is a familiar phenomenon, which we call Luddism, and resistance by management, or just the passive resistance of bureaucratic inertia, is a familiar complaint. Most businesses dismiss, disregard, and resent change, as well as unexpected opportunities for innovation. Most individuals do the same.

Because, as Schön says, "Innovation is something the corporate society must both espouse and resist" (78), the corporation develops a number of strategies for containing and controlling this necessary disruption. One of these strategies is the rationalized approach to managed innovation: "We suppress the surprising, uncertain, fuzzy, treacherous aspects of invention and innovation in the interest of this therapeutic view of them as clear, rational, and orderly" (77). In the realm of communication technologies, innovation is controlled and contained in part by the binding of tradition that occurs in the phenomenon of "remediation," described by Bolter and Grusin, and in what others have called the "replication" in new technologies of features familiar from older ones (for example, Shepherd and Watters; Yates, Orlikowski and Okamura).

Likewise rhetoric both espouses and resists innovation. Historically rhetoric has been understood as favoring the status quo, as a tool of the elite, and its lack of interest in innovation is what many believe made it seem irrelevant to the early modern period and the seventeenth-century scientific revolution. Rhetorical resources such as the commonplaces, generic conventions, and deference to the premises of the audience are all reservoirs of conservatism that resist or buffer change. Rhetorical invention was understood as "managerial" because it selects and deploys proofs already created by means other than rhetoric; managerial rhetorics are demand-driven "pull" theories. Contemporary rhetoric has tried to restore a balance by providing supply-side "push" theories, which require an understanding of invention as "generative." Robert Scott introduced this notion explicitly in 1967 and a few years later led a committee of the National Developmental Project on Rhetoric to characterize invention as "a productive human thrust into the unknown" (Scott; Scott et al. 229). The committee called for a "generative theory of rhetoric" that would help explain "the coming-to-be of the novel, the new, the 'invented'" (230). Richard McKeon has also offered a vision for rhetoric in a technological age that fully incorporates the pushmi-pullyu dynamic I have been exploring here: "It should be a rhetoric which relates form to matter, instrumentality to product, presentation to content, agent to audience, intention to reason." This rhetoric he called "an architectonic productive art," which is "an art of creativity, of judgment, of disposition, and of organization" (63). Such an art of rhetoric can be a worthy complement to the powerful arts of technology, as both arts push and pull us into our own future.

## Works Cited

Bitzer, Lloyd. "The Rhetorical Situation." *Philosophy and Rhetoric* 1 (1968): 1–14.

Bolter, Jay David, and Richard Grusin. "Remediation." *Configurations* 4.3 (1996): 311–58.

Bryant, Donald C. "Rhetoric: Its Functions and Its Scope." *Quarterly Journal of Speech* 39.4 (1953): 401–24.

Garsten, Bryan. *Saving Persuasion: A Defense of Rhetoric and Judgment.* Cambridge: Harvard University Press, 2006.

Gibson, James Jerome. *The Ecological Approach to Visual Perception.* 1979. Hillsdale, N.J.: LEA, 1986.

McKeon, Richard. "The Uses of Rhetoric in a Technological Age: Architectonic Productive Arts." *The Prospect of Rhetoric: Report of the National Development Project.* Eds. Lloyd F. Bitzer and Edwin Black. Englewoods Cliffs, N.J.: Prentice Hall, 1971. 44–63.

Schön, Donald. "The Fear of Innovation." *International Science and Technology* (November 1966): 70–78.

Scott, Robert L. "On Viewing Rhetoric as Epistemic." *Central States Speech Journal* 18 (1967): 9–17.

Scott, Robert L., James R. Andrews, Howard H. Martin, J. Richard McNally, William F. Nelson, Michael M. Osborn, Arthur L. Smith, and Harold Zyskind. "Report of the Committee on the Nature of Rhetorical Invention." *The Prospect of Rhetoric: Report of the National Developmental Project.* Ed. Lloyd F. Bitzer and Edwin Black. Englewood Cliffs, N.J.: Prentice Hall, 1971. 228–36.

Shepherd, Michael, and Carolyn Watters. "The Evolution of Cybergenres." *31st Hawaii International Conference on System Sciences.* Ed. Ralph H. Sprague, Jr. Maui: IEEE Computer Society Press, 1998. 97–109.

Vatz, Richard E. "The Myth of the Rhetorical Situation." *Philosophy and Rhetoric* 6 (1973): 154–61.

Winner, Langdon. "Do Artifacts Have Politics?" *Daedalus* 109.1 (1980): 121–36.

Yates, JoAnne, Wanda J. Orlikowski, and Kazuo Okamura. "Explicit and Implicit Structuring of Genres in Electronic Communication: Reinforcement and Change of Social Interaction." *Organization Science* 10.1 (1999): 83–103.

# Acknowledgments

The contributions of many can be found in the pages of this volume. I first want to thank the authors for their challenging and insightful essays. It is always a pleasure to work with people who can make real sense of the technological world around them. Their work is emblematic of the explanatory power of rhetoric in thoroughly modern times. Jim Denton, acquisitions editor at the University of South Carolina Press, expressed early interest in the project and provided valuable guidance and feedback at each and every stage. His support enabled the volume to find a wonderful home at the Press. Its staff, particularly Linda Fogle and Bill Adams, deserve special recognition for their effort. The reviewers read the manuscript carefully and offered good advice. I am grateful to my colleague Jack Selzer for commenting on a draft of the introduction and to Michael Faris for helping with proofreading and indexing. I also want to acknowledge the work of Rebecca Wilson Lundin; she served as my editorial assistant for the project. Rebecca helped me organize the introduction, respond to author drafts, and prepare the manuscript. Her contributions were unending and crucial to the volume. She has my deepest gratitude. Finally, I want to express my gratitude to the faculty and students in the Penn State rhetoric program—past and present.

# Introduction

Stuart A. Selber

The essays in this volume invite readers to consider the ways in which rhetorics and technologies relate to each other—and to numerous other aspects, both material and symbolic, of writing and communication situations. Key arguments have been organized and developed for people interested in a sustained treatment of these constantly shifting relationships, which have significant implications for both scholars and teachers in rhetorical studies.

Rhetorical activities have always taken place in technological contexts of one sort or another, whether a scriptorium, a traditional classroom, a state-of-the-art cybertorium, or other work space, private as well as public. Today those contexts have become ever more visible because they have multiplied in number and influence, ever more involved because they increasingly encompass literate activity, and ever more contested because they embody values and aspirations. For these reasons, technological contexts have moved toward the center of disciplinary conversations and encouraged people to think expansively and sometimes untraditionally about their practices and perspectives. Indeed, rhetoric scholarship on writing and communication technologies—print, digital, or otherwise—offers a range of perspectives that spans numerous sites of professional interest.

This diversity is a function of the seemingly inexorable move toward specialization that can now be found across the disciplines. In rhetorical studies, this move has resulted in collectives of researchers who have dedicated themselves to a rather different world of discovery, one that is relatively fast-paced, inevitably pragmatic, and thoroughly multimodal. An example here is computers and composition, a species of rhetoric that is particularly interested in the pedagogical dimensions of writing technologies. Another is technical communication, which focuses, in part, on the applications of communication technologies in nonacademic settings. It makes a certain sense, of course, to invest in the development of deep expertise and encourage inquiry that is sensitive and responsive to specific clusters of complex problems. But specialization can become an issue if it permits a discipline to compartmentalize concerns that permeate knowledge boundaries.

In both theoretical and practical terms, technology does not really function as a separate category or subcategory of consequence. It tends to infuse each and every area of the discipline, even under fairly narrow circumstances. In fact, it is difficult to imagine a rhetorical activity untouched by ongoing developments in writing and communication technologies. Their increasingly widespread integration into all facets of culture has encouraged scholars and teachers to reinterpret (yet again) the traditional canons of rhetoric. Invention strategies, for instance, now address powerful search capabilities and the ways in which database structures shape access to an intellectual landscape. Rhetorical education on arrangement no longer assumes a linear organizational pattern—or a patient reader, for that matter. More than occasionally, writers and communicators today anticipate reader control with modular hypertexts that can support multiple interpretive pathways and that can invite textual transformations and revisions.

Concerns about style (written and spoken) have evolved in several different directions, especially since the convergence of Internet and broadcast technologies, which initiated an increase in the number and character of forums—official and unofficial, popular and scholarly—for discursive interactions. Yet even a simple medium like electronic mail or instant messaging can confound understandings of style. Although exchanges in these environments are, at least for the moment, primarily written, such exchanges mix markers of oral and literate performance in ways that either fascinate or distress those who study the evolving nature of language use. Some researchers emphasize the opportunity to witness emergent communication practices; others worry that the vernacular complexion of those practices contributes to the decline of literacy.

The canon of memory has also been reinterpreted to accommodate current literacy concerns. In oral traditions, this canon concentrated to a great extent on strategies for recall and recognition, strategies that undoubtedly became less and less important to the literate activities of print cultures. These days, however, memory has been resuscitated by the brain-as-computer metaphor, which informs the assumptions of innumerable investigations of information processing, including projects working to understand the nature of writing and reading in a liminal moment marking the shift from analog to digital systems. In addition, by constantly mapping movement in and across networks, the Internet itself has become a robust storehouse for both individual and collective memory. The issues here have profound social dimensions and reach beyond functional worries over user disorientation and information overload, over recall and recognition in large-scale information spaces. Who controls the memory storehouses constructed by users as they navigate the Internet? Who has access to the data? To what purposes is it put? These are just a few of the questions that readily come to mind.

Updated notions of delivery involve a comparable trajectory of functional and social development. Since at least the 1980s, computer programs have enabled people to assume increasing control over the structural and presentational aspects of texts. If ancient rhetoricians focused on areas such as gesture, tone, and expression for oral occasions, modern rhetoricians attend to numerous design elements and complex relationships between word and image, moving as well as still, and between word or image (or both) and sound. Although the printing press ushered in a shift from script to print, making information available to a much larger segment of the population, it did not immediately encourage writers and communicators to become designers of their texts. Nor did it immediately redistribute control over the production and circulation of texts: the printing press enabled one-to-many communication, but early communicators were typically from elite classes, and their messages were typically associated with official institutions like churches and governments. In contrast, people nowadays have unprecedented access to powerful systems for publishing and information distribution. This access has encouraged computer users to become producers/designers of both print-destined and online texts. Readers, too, have experienced an expanded role, for the ways in which they configure their personal computing devices affect the look and feel of online texts. More than ever before, perhaps, issues of delivery intertwine writers, readers, technologies, and cultures in new and intimate ways.

In spatial terms, the previous discussion stresses memory and delivery over invention, arrangement, and style, not because they are more vital or important, but because contemporary literacy contexts have brought added significance and new meaning to the final two canons of rhetoric, which historically have been somewhat neglected. This situation is emblematic of the recursive relationship between rhetorics and technologies. Various rhetorical frameworks have been mobilized to help explain design and use contexts for writing and communication technologies. Those contexts, in turn, can influence how people think about the nature and role of rhetorical theory and practice. Early discussions of distance education, for example, advised teachers to transfer to the screen their pedagogical approaches from the brick-and-mortar classroom. Teachers new to computer-based environments were especially encouraged to foreground rhetorical learning objectives and desired outcomes. Although learning objectives can transcend media boundaries, such advice assumed that computers were more or less neutral, that technologies and their contexts would not shape or challenge how teachers thought about domain content or about the art and craft of teaching itself. This assumption was useful to new teachers, but a nondialogic perspective failed to emphasize the interanimations of rhetorics and technologies.

The assumptions informing this volume exhibit greater social sensitivity. First, technological contexts encompass more than just physical devices like books and computers. They also include systems, techniques, and methods for rationalizing work and society. After all, language itself is something of a technology. Second, technological contexts are, in a very real sense, overdetermined: multiple forces and factors shape the directions and priorities of technological projects. In other words, there is no one-to-one correspondence between technology and change, innovation, or social transformation. Third, and perhaps ironically, technological contexts entail human as well as technical problems, problems of subjectivity, identity, agency, materiality, methodology, pedagogy, representation, and more. As these assumptions suggest, technological contexts are decidedly rhetorical in character.

This reality provides the conceptual backdrop for a series of questions motivating the essays in this volume: What might history contribute to a rhetorical understanding of technological contexts writ broadly? How does rhetoric, as it has been traditionally mapped out, both illuminate and fail to illuminate the design and use of literacy technologies? How do issues of technology intersect with issues of identity, subjectivity, and agency? With race, class, gender, and ability? With other contemporary theory issues and categories? What do productive technologies look like in terms of their design? What specific contributions can rhetoricians hope to make to technological design practices? How are people currently working with technologies of production and reception? What, then, does it now mean to read and write? Teach and learn? Communicate and collaborate with others? What types of challenges accompany the task of integrating technologies into institutions? What are the ethical and professional questions raised by technology and its current contexts? How should rhetorical studies think about such matters? What might be especially productive methods for studying and evaluating technology in context? On the whole, such questions suggest a rich landscape for exploration and analysis.

This landscape is mapped into three sections by the volume: "Redrawing Borders and Boundaries," "Constructing Discourses and Communities," and "Understanding Writing and Communication Practices." The unfolding logic of these sections draws on and interrogates familiar categories and explanations, creating productive tensions that provoke critical reflection on the impact and state of rhetorical studies. Furthermore, the macrostructure captures the diversity of argumentative and research work in the discipline. This diversity is important, intentional, and an artifact of authors who recognize the rhetorical vectors of research practices. Grounded theory, discourse and artifact analysis, participant observation, case study, historical reasoning, and interdisciplinary inquiry—the authors employ these and other approaches to comment differently on the evolution and complexity of the rhetoric-technology nexus. Although the authors mobilize a variety of methods and arguments, each attempts to engage a significant problem context in both useful and complicated ways. The volume integrates practice and theory at fundamental levels—approaches, assumptions, questions—to ensure relevance for scholars and teachers in rhetoric.

The problem contexts scaffold issues in several discernible patterns of coherence. The primary pattern suggests a model for rhetorical education, one that begins with new approaches to writing and communication and ends with challenges to enacting productive change. In between, the essays disclose an array of literacy complexities that can have an effect on both individuals and the discipline. This arrangement is useful because it highlights both the ubiquity and entanglements of technology in a contemporary society. Students new to rhetorical studies, especially at the graduate level, often start with the ancients and with approaches to discourse that have been studied (and modified) for centuries. Technology, in a standard pedagogical approach, is either ignored or treated as an add-on to rhetorical thinking and conceptualization. Those who ignore technology hide beyond the insights of the past to reject new configurations of rhetoric. Those who picture technology as an add-on underestimate the extent to which dialectic tensions occupy the literate spaces and activities of a digital age. Authors in this volume introduce a genre of scholarship that blends rhetorical criticism and technology studies, bridging the past and future without self-consciousness. The essays serve as exemplars of this new genre and bring consistent attention to issues and relationships that can no longer be avoided or left implicit in conceptualizations of rhetoric that guide education and work.

In addition, the section and essay titles themselves signal keywords in modern rhetorical studies. In the language of new media and Internet-based communication, these keywords constitute a "tag cloud" of associations (without the user-generated weightings one finds on the Internet), reflecting lines of articulated discussion in the volume. These lines are conversation starters for readers. There are also thematic connections established through the repetition of certain critical practices, which are distinctive features of the essay genre in this volume. Authors do not see technology users as detached from operations that generate and legitimate knowledge. All of the essays consider human effects and interventions in technological contexts. They also assume a postcritical intellectual stance, meaning that technology is understood to be either an intrinsic or inescapable aspect of culture, an aspect that should be dealt with directly, seriously, and productively. Such realism is an essential component of any agenda claiming to address matters of significance to rhetorical studies.

Part 1, "Redrawing Borders and Boundaries," offers new ways to think about contexts and traditions for writing and communication that have become foundational (and indispensable) to disciplinary development. Authors in this section interrogate technologized aspects of discourse production, reception, and dissemination, focusing on what it means to participate in a rhetorical milieu that integrates writers and writing systems (somewhat) seamlessly and dialectically and that involves both conventional and emergent communication practices. In the opening essay, "Being Linked to the Matrix: Biology, Technology, and Writing," Marilyn M. Cooper reimagines the basic structures of the rhetorical situation, offering a view that embraces dynamic interaction, negotiation, and coordination as major elements. Her approach, which draws on phenomenology and complexity theory, departs from referential perspectives on language development and use—perspectives that tend to stress rationality and the intellectual autonomy of writers and communicators. It also deviates from social perspectives that either diminish the role of embodiment in cognition and communication or fail to account for interactions that motivate people to engage in discursive practices. The approach offered by Cooper understands writing as an ongoing process of responding to others, to texts, to contexts: a many-sided, contingent process that is not so much a function of intended plans or actions but of coordinating and mobilizing resources in settings of relevance to writers and communicators. The meanings and uses of technology, in this articulation of the contours of rhetoric, emerge from human interactions and are part of the cognitive ecologies elaborated by humans in acts of literacy.

The discourse process described by Cooper is not exclusive to invention and production activities, however. Nor is it always initiated and advanced by human agents. People can now be written by the very technologies that they employ, particularly in their roles as readers and consumers of online information. This recursive event happens, unwittingly or otherwise, in even mundane use situations. "Among Texts," by Johndan Johnson-Eilola, considers the human-machine feedback circuits that have been promoted by the development of networked spaces for literacy and education. According to Johnson-Eilola, texts have become an active component of database systems, both personal and large-scale, functioning as nodes on digital networks that not only provide information to readers but also generate information about readers. No longer physically discrete artifacts, texts today can record literacy habits, activities, and experiences. They can even communicate with one another and respond to readers. Johnson-Eilola discusses these expanded capabilities and speculates on a future in which metadata about texts is central to the rhetorical situation.

Redrawing disciplinary landscapes to accommodate other boundaries and practices is an indeterminate enterprise—an unending series of negotiations, really, amid forces (and cycles) of reform and stasis. As Johnson-Eilola demonstrates in his social history of texts, any number of meanings can be attached to a literacy device or activity. In response to the confines of received traditions and the domesticated meanings they can encourage, scholars and teachers have begun to generate alternative perspectives on textual production and use. Not surprisingly, these other views have emerged from a confluence of historical and societal factors. In "Serial Composition," Geoffrey Sirc articulates a vision for rhetorical studies that promises to evolve the field in a productive direction. The factors discussed by Sirc—the rise of minimalism in the art world, the publication of the first major study of composition instruction in the modern era, and the development of the compact audiocassette—all have roots in the same historical moment, 1963, a moment that anticipates contemporary literacy issues and provides antecedents for disciplinary approaches organized around serial versus combinatory logics. The combinatory logic of rhetoric expresses itself most directly (but perhaps too complicatedly) in the thesis-driven essay, which is a hallmark of rhetorical education. Although Sirc does not dismiss the usefulness of instruction that focuses on persuasion, coherence, and deductive organizational patterns, such priorities can lack explanatory power in literacy contexts that mix and remix genres, integrate multiple media, and provide access to and control over an expanded set of literacy resources. The approach Sirc offers is inspired by composition practices that permit organizational structures to surface from material and practical engagements with texts and their contexts. The serial form of his approach, which is instantiated on the Internet by the grammar of MP3 Weblogs, values juxtapositions and associations rather than causal relationships and intricate discursive arrangements. This essay, in the end, stands as a model for rhetorical development that is sensitive to the realities of digital media and to the social-historical forces that can maintain status quo ideas in the profession.

Part 2, "Constructing Discourses and Communities," examines the signifying practices of people writing about, deploying, and structuring rhetorical activity in technological contexts. The discourses of these varied audiences—scholars, users, and designers—contribute considerably to the problems and prospects for online communication and for communication about online environments. In "Appeals to the Body in Eco-Rhetoric and Techno-Rhetoric," M. Jimmie Killingsworth argues that the professional discourse on techno-rhetoric can be too inattentive to the material dimensions of computer-based work. More specifically, he worries about the impulses of scholars and researchers whose formulations of cyberspace neglect the body and its needs or diminish connections between bodies and other physical and social contexts. Using the discourse of eco-rhetoric as a contrast point, Killingsworth identifies instructive moments in the scholarly literature in which people fail to account for the body as an integral part of the brain-computer interface. He then points to practical and theoretical consequences of this neglect for literacy and learning and for academic programs in writing and communication.

If scholarly discourses on technology can fail to account adequately for the materiality of the body, they can also neglect its perceptual and sensual operations. Such neglectful treatment, especially in the context of new media and digital communication, has implications for rhetorical studies that are not to be minimized or dismissed. In "Unfitting Beauties of Transducing Bodies," Anne Frances Wysocki argues that the field should begin to cultivate conceptions of aesthetics and perception that recognize the unique characteristics of new media projects. The characteristics in which Wysocki is interested engage bodily senses in novel or unexpected ways. But they do more than just that. In addition to mediating interpretive processes, the senses, in an age of interactive digital technologies, can also function as a literal component of new media projects, helping to constitute the experiences of users and the expressions of artifacts and interfaces. This expansive design practice, however, has not always been accompanied by conceptual adjustments to aesthetic or perceptive theories, adjustments that are warranted by the interactive possibilities of online texts. As Wysocki notes, discussions of new media projects tend to count on origin stories for the appearance of aesthetics as a named field. On one hand, these stories help to legitimate such projects by evoking traditional criteria and definitions. On the other hand, they assume that sense experience is a private, natural phenomenon rather than a complex social process inflected by cultural forces. This belief system can lead people to underestimate the extent to which the sensuous histories of users influence their perceptions of new media projects. It can also discourage ethical deliberations over the imprint of cultural forces on sensuous histories.

Scholarly discourses, then, are a potent delivery system for representations of technology that inform the work of the profession. But it is also instructive to pay attention to the discourse practices of users, because concrete rhetorical gestures situate technology in time and space and reflect motivations, needs, and values that can shape the nature of digital environments. "The Rhetorics of Online Autism Advocacy: A Case for Rhetorical Listening" analyzes discourses about autism and autistics in Internet forums that function as alternatives to mass media outlets and their sources of information. Authors Paul Heilker and Jason King evaluate characterizations of autism in the mainstream media, noting, above all, that autistic communities rarely (if ever) have opportunities to speak for themselves, that others—celebrities, parents, journalists—regulate dominant discourses about autism in the public sphere, discourses that promote a disease model of disability and account for only a portion of the issues with which autistics are concerned. In response to this situation, higher-functioning, verbal autistics have begun to use the Internet to advance their own voices and agendas. Heilker and King discuss technical affordances and social dynamics in such efforts and encourage a rhetoric of reconciliation that might enable productive online dialogue between those with conflicting views of autism.

The discourses of users and scholars inevitably come into contact with the discourses of designers, who condition and configure how people think about both technology and their tasks. Although designer discourses can be overridden and rewritten, they define terms of engagement and help to structure the spaces in which literate activity is crafted and enacted. For these reasons, designers of online spaces should be rhetorically sensitive as well as technically astute.

In "Narrating the Future: Scenarios and the Cult of Specification," John M. Carroll explains and illustrates an approach to design that is conscious of the provisional character of communication. His philosophy recognizes the problem-solution cycle permeating technological contexts: new technologies both solve and create human problems, and therefore necessitate more design work and invention. This cycle is generative, endless, and essentially a function of how people interpret design representations. To promote a rhetorical perspective, Carroll asks designers to confront the problem-solution cycle with heuristic scenarios that can help them imagine and anticipate the complexities of user settings. These scenarios emphasize shifting circumstances, foreground contextual details, and stimulate design deliberations. Scenarios, for Carroll, stand out against formal specifications, which are top-down, machine-oriented descriptions of artifacts in development. Although specifications provide useful information, scenarios address the concerns of rhetoric, in part by creating meaningful dialogue between user and designer discourses and accounting for contingencies in communication events.

Part 3, "Understanding Writing and Communication Practices," maps transformations in thinking about invention, production, literacy, and power that have been forged at the intersections of rhetorics and technologies. Authors in this section discuss projects and endeavors that illuminate the ways in which discourse activities can evolve to reflect emerging socio-political realities, technologies, and educational issues. "Technology, Genre, and Gender: The Case of Power Structure Research," by Susan Wells, discusses ways in which subaltern groups leveraged the capabilities of emergent print technologies in order to critique the assumptions and directions of received institutions. Wells conducts visual-rhetorical analyses of underground publications from various social movements, showing how new techniques for graphic design created a vernacular style that helped cultures of resistance achieve political and social aims. The success of these publication efforts relied on the participatory practices of design amateurs who reinvented the genres of conventional journalism and invented genres that mediated between academic and movement audiences. These design amateurs, in addition, saw themselves as issue advocates rather than objective reporters. Their transparent stances contributed to an ethos that was constituted, at least in part, through publications that relied less on the authority of position and status and more on the authority of engagement and direct experience. Wells, however, does not limit her focus to history. At the end of the essay, she relates past contexts to the present in a manner that helps rhetoricians think (once again) about the nature of new media.

As Wells shows, settings for literacy can supply opportunities for cultural intervention. Such interventions, especially in the context of Internet-based communication, often involve alternative views of interaction and exchange. To begin his essay, "Rhetoric in (as) a Digital Economy," James E. Porter urges scholars and researchers to be alert to the economics of rhetoric in social networking environments, an economics that occupies two related dimensions: non-monetary value that (nevertheless) encourages active, ambitious participation on the Internet; and reproduction and distribution mechanisms for multimodal texts that are inexpensive or even free. These two dimensions, according to Porter, have the potential to recast the roles of audiences and writers, particularly in situations involving complex social problems or idiosyncratic end-user problems. In such situations, the wisdom of the crowd often proves to be more valuable than the wisdom of expert writers. Consequently writers and communicators must learn to support audiences who are beginning to blur the lines between content consumers and content producers. As Porter demonstrates, building and managing support systems for user-generated content requires social and political understandings of networked-based writing and reading and a revived—and revised—notion of the canon of delivery.

New media users also collaborate on other sorts of intellectual problems, including attempting to understand and take part in large-scale narrative experiences that encompass multiple media and span various time-space frames. In "Literate Acts in Convergence Culture: *Lost* as Transmedia Narrative," Debra Journet investigates the logistical and interpretive challenges presented by media spaces that combine and remix conventions from video games and television shows, creating new modes of engagement in which audiences have a direct say in the unfolding directions of both major and minor plot stories. Her analysis of collective intelligence in one participatory culture reveals an impressive level of audience engagement with imaginative texts that any rhetoric scholar or teacher would welcome and appreciate. Although this engagement consists of recognizable literacy practices—for example, identifying intertextual references and debating authorial intentions—it also includes new literate challenges associated with retrieving and sharing online information, navigating and integrating multiple media, making meaning in distributed collaborative spaces, and recognizing how media and mode shape narrative structures. As these (and other) challenges suggest, texts in a convergence culture are meant to be operated and not just read, serving as interactive prompts that urge people to participate in the construction and reconstruction of richly resourced narrative experiences.

Technological contexts raise important, difficult, and complex questions that encompass the five canons of rhetoric and numerous other concepts and concerns. In response to the wide scope and considerable reach of these multilayered questions, authors in this volume work both with and against the grain of disciplinary thought to illuminate the existent and emergent intersections of rhetorics and technologies. They also employ a panoply of approaches and arguments that attempt to refashion disciplinary thought that is not receptive to the literacy practices of today or to the projected realities of tomorrow. In addition, each essay models productive habits of mind for technological contexts, situating technologies in space, time, and culture and acknowledging the ways in which technologies, like rhetorics themselves, serve as interfaces for human relations and endeavors. Such habits of mind engender meaningful conversations about technology and clarify the stakes of technological projects not only for rhetorical studies but also for society at large. This volume begins to show that the stakes for everyone are very high and very real indeed.

# 1

# Redrawing Borders and Boundaries

# Being Linked to the Matrix

Biology, Technology, and Writing

Marilyn M. Cooper

Edward Hoagland writes:

> I'd lie on my back on a patch of moss watching a swaying poplar's branches interlace with another's, and the tremulous leaves vibrate, and the clouds forgather to parade zoologically overhead, and felt linked to the whole matrix, as you either do or you don't through the rest of your life. And childhood—nine or ten, I think—is when this best happens. It's when you develop a capacity for quiet, a confidence in your solitude, your rapport with a Nature both animate and not much so: what winged things possibly feel, the blessing of water, the rhythm of weather, and what might bite you and what will not. (49–50)

Perhaps it was because my father is a fisheries biologist and we spent a lot of time on lakes when I was a child that I know this feeling of being linked to the whole matrix and that it is deeply sedimented into my thinking about all aspects of life. In 1986 I published an article about the ecology of writing in which I struggled to articulate my sense that writing is social action, not simply an activity that takes place in a social context. I hoped to encourage a view of writing and writers as fully engaged in social practices: I wanted to emphasize how writers and writings shape their social environments and are shaped by them in a manner analogous to the way organisms interact with their environments. It was also around 1986 that some of the classic expositions of complex systems theory were coming out: the first English edition of Benoît Mandelbrot's *The Fractal Geometry of Nature* (1982), Ilya Prigogine and Isabelle Stengers's *Order Out of Chaos* (1984), Terry Winograd and Fernando Flores's *Understanding Computers and Cognition* (1986), and Humberto Maturana and Francisco Varela's *The Tree of Knowledge* (1987). Although I did not read these works until much later, I was aware of chaos theory (through the popular account *Chaos: Making a New Science*, by James Gleick, published in 1987), and after the publication of my article I increasingly thought that the systems of writing are not just analogous to ecological systems but are driven by the same principles. In meetings at the Los Alamos lab, the Santa Fe Institute, MIT, and the University of Illinois, researchers in the diverse fields of economics, physics, biology, cybernetics, mathematics, and meteorology were coming to similar conclusions. The challenges of investigating chaotic phenomena blurred the boundaries between disciplines and between the realms they studied. I, too, believed that once social phenomena such as writing were viewed as complex systems, the distinction between nature and culture would come into question, just as the related dualities of mind and body, subjectivity and objectivity, had. When I read about autopoetic systems, I felt as though someone was explaining something I implicitly knew, or, as Merleau-Ponty says about reading Husserl or Heidegger, as though I was "recognizing what [I] had been waiting for" (viii). When I finally read Gregory Bateson's *Steps to an Ecology of Mind* (which I had owned, unread, almost since it was published in 1972) and Maturana and Varela's *The Tree of Knowledge*, I found my nascent vision of writing as a web clarified and expanded in Bateson's idea of mind as "immanent in the total interconnected social system and planetary ecology" (461) and in Maturana and Varela's claim that "we humans, as humans, exist in the network of structural couplings that we continually weave through the permanent linguistic trophallaxis of our behavior" (234).

The study of complexity, originating around the early twentieth century,[1] has grown exponentially in the past three decades. Proceeding under various titles—chaos, complexity, emergence, autopoesis, self-organization—this research has generated new understandings of, among other things, brains, fractals, thermodynamics, ontogeny, software, ecosystems, synchronicity, and communication and economic systems. It is also drawn on increasingly in theories of virtual humanism and network culture, as in the work of Mark Hansen, Katherine Hayles, and Mark Taylor. At the same time, cognate understandings of systems developed in the work of phenomenologists Heidegger and Merleau-Ponty, as well as in related theories of Bergson, Wittgenstein, Deleuze and Guattari, Bourdieu, Giddens, Latour, and Bakhtin, as well as Edwin Hutchins in distributed cognition and Jean Lave and Etienne Wenger in situated learning, among others.[2] My reading in all of these theories (see fig. 1.1) increasingly led me to the conclusion that writing is not a matter of autonomously intended action on the world, but more like monitoring, nudging, adapting, adjusting—in short, responding to the world.

Although few researchers have explicitly applied complexity theory to writing,[3] several groups of researchers in rhetoric, composition, and literacy studies, working from some of the related theories just mentioned, have offered understandings of writing as a dynamic and interactive system. Members of the New

Fig. 1.1 Source theories

London Group, many of whom have backgrounds in social theories of meaning and language, envision language systems as structured by the interactions of users; compositionists such as Charles Bazerman, David Russell, and Paul Prior draw on activity theory to conceive of writing as a means of making and transforming social worlds; and Victor Vitanza and his "third sophistic" followers derive a vision of writing as an embodied and open system from their readings of Nietzsche and Deleuze and Guattari. My approach to writing differs from theirs primarily in my emphasis on writing as arising from responses to others and to social and physical environments, responses that involve both body and mind and are only partly and sometimes intentional.

The New London Group and activity theorists both see writing as predominantly an intentional cognitive process. The New London Group acknowledges that "the human mind is embodied, situated, and social" (30) and that "immersion in a community of learners" is necessary for learning literate practices (31), but they argue that "conscious control and awareness of what one knows and does" and the ability to "critique what they are learning" are crucial to the activity of writing (32). Activity theory recognizes a large role for tacit consciousness in writing, but activity theorists, like the New London Group, focus on writing as a conscious cognitive process. Bazerman and Russell argue, "Things human exist in an evanescent world held up by focused consciousness and attention and activity" (1). The emphasis on consciousness in activity theory can be traced to the Marxian assertion of fundamental differences between humans and other animals. Vassily Davydov explains, "In the process of human anthropogenesis, a break occurred between organic needs and the means of satisfying them, that is, human beings lost their instincts" (49). In place of instincts, humans use social forms of activity to satisfy their needs, and consciousness supplies the "internal images" that link need and goal: it is "people's ideal images that make it possible to foresee the product" (40). But theories and studies of cognition in animals, beginning in the early twentieth century with Uexküll and continuing to the present (see Csányi, Griffin, and Hauser), as well as contemporary studies of human cognition, have undermined any notion that humans have lost their instincts or that the link between need and goal is determined by ideal internal images.

Vitanza, in contrast, seems to envision writers as merely channeling writing. His emphasis on the fluidity—the flow—of writing turns it into an autonomous Nietzschean life force that animates human agents who are largely or entirely unconscious of its desires. Writers are not understood to be making choices, but are driven to write. Vitanza says, **"What writing . . . wants is a writer! . . . A body filled with tics that cannot but (not) write!"** (4), a statement highly reminiscent of Barthes's idea of the death of the author—"the author is never more than the instance writing" (145). Although Vitanza's understanding of writing as engaging the body is a good corrective to the idea that writing is dominantly cognitive, he seems to acknowledge no role for intentional response.

In this essay I argue that writing is an embodied interaction with other beings and our environments. As a result, writing is as much a biological as a cultural practice: the practices that are writing emerge as people respond to others and to their world; they are not the product of minds somehow separated from bodies nor of innate technical or linguistic abilities. Furthermore, I argue that writing and technology are cognate practices. Arising as an epiphenomenon of engaged action in the world, tools and words play the same role in our lives. As concrete objects that can be manipulated and can store information, tools and words extend cognitive processes beyond the individual brain. Other beings can also be recruited in the same way, as dogs extend the abilities of shepherds to control sheep and editors extend the abilities of writers to consider other perspectives. As I use the term, writing often describes both linguistic and technological practices, practices that function to elaborate cognitive ecologies such as those that make sheep herding and publishing possible. Writing in this sense is what makes us human. The extent of our abilities to elaborate cognitive ecologies may set us apart from other animals, but no nonbiological source is needed to account for our abilities, an argument buttressed by evidence that other animals share many of the same abilities. Neither language nor technology is foreign to our nature; tools and words are us, not things we create and use.

To get a sense of what writing looks like from this perspective, consider these examples. First, the use of DEVONthink by professional writer Steven Johnson. DEVONthink is sophisticated indexing software that works on an archive of the writer's writings and notes and excerpts from the writer's reading; Johnson explains that it not only searches on specific words but also "learns associations between individual words, by tracking the frequency with which words appear near each other." He describes how in working on a book project involving the history of the London sewers, he ran a search on "sewage." Among the results, he also received references to "waste," a word that often occurs with "sewage," including a quote about how calcium waste products were repurposed into bones in the evolution of vertebrates.

> That might seem like an errant result, but it sent me off on a long and fruitful tangent into the way complex systems—whether cities or bodies—find productive uses for the waste they create. It's still early, but I may well get an entire chapter out of that little spark of an idea.
>
> Now, strictly speaking, who is responsible for that initial idea? Was it me or the software? It sounds like a facetious question, but I mean it seriously. Obviously the computer wasn't conscious of the idea taking shape, and I supplied the conceptual glue that linked the London sewers to cell metabolism. But I'm not at all confident I would have made the initial connection without the help of the software. The idea was a true collaboration, two very different kinds of intelligence playing off each other, one carbon-based, the other silicon.

Here is a second example. A group of students in a writing class make a documentary video reporting their research on the Paulding light, a well-known mystery in the Upper Peninsula of Michigan. Their research involves a trip to the site, where, using their cell phones and a GPS unit, they establish that the light comes from headlights on a highway in the distance, and they use a video camera to record their observations and commentary. Although their teacher might be tempted to exclaim at their cleverness in using all that technology, for them the cell phones, GPS unit, and video camera simply come to hand as part of the already-established consensual domain of these extensively mediated and technologized students, students for whom nearly continual communication with others, never being lost, and being immersed in images of their own and others' making has been a way of life. Through their actions, extended in their prosthetic technologies, they create the world as knowable, a world in which there are no obstacles to ascertaining precise positions and exchanging words and images and in which, as a consequence, there are no mysteries.

These examples illustrate three points I want to make about writing considered as a biological/cultural, linguistic/technological practice. First, notice that in the process of writing, words and tools do not normally arise as separate objects to be used but are experienced as part of our bodies and brains; they are, as Heidegger says, ready-to-hand, not present-at-hand. Steven Johnson experiences the genesis of the idea of how complex systems repurpose wastes as a collaboration between him and the indexing system, a productive interaction between carbon- and silicon-based intelligences; and in both examples, the technologies of DEVONthink, GPS unit, cell phones, and video camera are as much a part of the writers as their hands and eyes. Second, writing is not just autonomous social action but always an interaction with other beings and objects in our surroundings, an ongoing process of stimulus and response that we habitually misconceive as autonomous planned action. The "errant result" returned by DEVONthink stimulated Johnson to think in different ways about waste, and the teacher in the second example credits the students with bright ideas about using technology to achieve their goals that probably never entered their conscious minds. Third, writing is a complex system organized by dense interactions of writers and their worlds. DEVONthink, like all of the technologies of propagating and indexing writing, amplifies and makes visible these dense interactions out of which invention arises. And the students "use" "communication" technologies to investigate the Paulding light because through interactions with one another and their worlds they have become habituated to how these technologies abolish distance, and they thus experience (and expect) all people and places to be always accessible.

## Words and Tools Arise from Interaction

Neither words nor tools exist prior to or separately from human action. They arise as an epiphenomenon of that action and are continually reconfigured or reinterpreted as they arise again in different situations. Perhaps the best-known enunciation of this understanding of language is Wittgenstein's in *Philosophical Investigations*. Disputing Augustine's contention that the meaning of a word is what it stands for, Wittgenstein argues that the meanings of words arise from their use in social interaction. He imagines a simple protolanguage used by a builder and his assistant: "A is building with building-stones: there are blocks, pillars, slabs, and beams. B has to pass the stones, and that in the order in which A needs them. For this purpose they use a language consisting of the words 'block,' 'pillar,' 'slab,' 'beam.' A calls them out;—B brings the stone which he has learnt to bring at such-and-such a call" (sec. 2).

Wittgenstein argues that this language is learned by training B to respond to the words in particular ways. He asks, "Don't you understand the call 'Slab!' if you act upon it in such-and-such a way?" (sec. 6). B has learned the use of the words and that is all he needs to know to understand and play this language game. Wittgenstein asks, "Now what do the words of this language *signify?*" and answers with another question, "What is supposed to shew what they signify, if not the kind of use they have?" (sec. 10). He concludes that "the *speaking* of language is part of an activity, or of a form of life" (sec. 23).

Paleoanthropologist Alison Wray advances a similar theory of the origin of language in what she calls a protolanguage that consisted of holistic utterances that later developed into referential symbols. More along the lines of Wittgenstein, another anthropologist, Tim Ingold, argues that in use, language never "advances" to the level of the referential. Drawing on Merleau-Ponty's idea that "there are no conventional signs, . . . there are only words into which the history of a whole language has been compressed" (qtd. in Ingold, "Tool-Use" 435), Ingold points out that writers are always immersed in a meaningful relational world and that far from being founded in convention, *words gather their meanings from the relational properties of the world itself* " (Ingold, *Perception* 409).

Animal behaviorists observe that animal communication also seems to emerge from interactions with others and with their surroundings rather than from acts of referring. A much-examined example is the alarm calls of the vervet monkeys in Africa. Vervet monkeys give one of three types of calls when they see one of their three main predators—leopards, eagles, and snakes—and, like Wittgenstein's builder's assistant, the monkeys respond to these calls with different kinds of behavior. In response to the leopard call, they climb into trees; in response to the eagle call, they dive into bushes; and in response to the snake call, they stand up and look around (Griffin 158). Thus animal behaviorists conclude that the calls might better be translated in behavioral rather than referential terms: for example, "behave in a way to escape a leopard," rather than "there's a leopard."

Chilean biologist Humberto Maturana elaborates these ideas in his argument that language is not a symbolic system or an instrument of communication but a result of the coordination of behavior. Conceiving of language as an instrument of communication is misleading in that it mistakes the result for the cause, as he explains: "Human beings can talk about things because they generate the things they talk about by talking about them. That is, human beings can talk about things because they generate them by making distinctions that specify them in a consensual domain" ("Biology" 56). What Wittgenstein calls language games Maturana calls "the flow of coordination of behaviors" ("Nature" 462) that results in the establishment of a consensual domain, a taken-for-granted world that arises in the interaction. He offers an example of a woman hailing a taxi by meeting the gaze of a taxi driver and making a circular hand gesture, a learned coordination of the behaviors of getting attention and committing to hiring. In this coordination of behaviors, the taxi arises as a means of transportation. In other domains the taxi may arise as something else—an art object, say—or may not be individually distinguished—if it arises as part of a traffic jam, for instance. Such domains of interobjectivity are consensual not in the sense of being agreed upon, but in the sense of a "coherent transformation of behavior of two or more organisms as they live together, [which] occurs as an unintended result of that living together" ("Nature" 463). In sum, Maturana says, "We literally create the world in which we live by living it" ("Biology" 61).

Ingold extends this argument to tools. He contends that "tools—like words—are used to mediate an active engagement with the environment rather than to assert control over it. Meaning, thus, is not imposed on the world but arises out of that engagement" ("Tool-Use" 433). Just as Wittgenstein and Maturana see the meaning of words as arising in their use, Ingold sees the purpose of tools as arising in their use. He says, "An object—it could be a stone or a piece of wood—*becomes* a tool through becoming conjoined to a technique. . . . Thus the tool is not a mere mechanical adjunct to the body, serving to deliver a set of commands issued to it by the mind; rather it extends the whole person" ("Tool-Use" 440). Consider again in this context how indexing tools like DEVONthink enable a writer's invention and how the notion of an index itself arose out of the practice of collating instances of sign use.

Thirty years earlier, in 1964, French paleoanthropologist André Leroi-Gourhan made much the same observation about tool use by early human ancestors: "We perceive our intelligence as being a single entity and our tools as the noble fruit of our thought, whereas the Australanthropians, by contrast, seem to have possessed their tools in much the same way as an animal has claws. They appear to have acquired them, not through some flash of genius which, one fine day, led them to pick up a sharp-edged pebble and use it as an extension of their fist (an infantile hypothesis well-beloved of many works of popularization), but as if their brains and their bodies had gradually exuded them" (106). The discovery of tool manufacture and use in human ancestors who had yet to acquire the proportionally giant brains of *Homo sapiens* argues against the development of technology as a conscious mental achievement, a matter of inventing a tool for a particular use. Instead, tools seem to have arisen out of physical and kinetic coordinations between agents and their environment—they result from actions of shaping rather than being instruments designed for shaping.

### Writing Is Interaction

Writing is always an interaction with other beings and objects in our surroundings, an interaction that we habitually misconceive as autonomous action that begins in our minds. The idea that words and tools are "the noble fruit of our thought," invented to serve a specific purpose, is a correlate to our tendency to interpret our ideas for book chapters or production of a video research report as the result of intentions and plans arising autonomously in our minds rather than as arising from interactions with our surroundings. According to this view, all we need to create texts is linguistic, rhetorical, and technological cognitive abilities. Tim Ingold targets this idea, "the assumption that for people to speak they must first 'have' language, or for people to use tools they must first 'have' technology—or indeed for people to engage in intelligent activities of any kind they must first 'have' intelligence" *(Perception* 407), as preventing us from understanding that "skill . . . is a property not of the individual human body as a biophysical entity, a thing-in-itself, but of the total field of relations constituted by the presence of the organism-person, indissolubly body and mind, in a richly structured environment" (353).

That skill is based in intelligence is such a common assumption that the headline on a report of tool use among New Caledonian crows in USA Today (Friend) reads: "Crows exceed expected intelligence levels," a conclusion not really borne out in the story. In experiments conducted at Oxford University, Betty, a female crow, and Abel, a bigger, dominant male, were faced with a piece of meat in a tube and given a hook and a straight piece of wire; both crows quickly chose the hook and used it to get the meat. More interesting, when Abel stole Betty's hook, she made a hook out of a straight wire and continued to get food. When retested with just straight wires, she made a hook nine out of ten times. Experimenters did not test Abel for this ability, because "dominant males employ more efficient, though perhaps less clever strategies: They wait until the work is done and steal the food from subordinates." Betty and Abel achieve these feats not because they have technological or social intelligence but by interacting with their surroundings in ways that benefit them. Consider again, in this connection, how the students created the documentary video about the Paulding light by interacting with one another in their extensively mediated and technologized environment.

As examples of skill as a matter of organisms interacting with their surroundings, Ingold uses the weaving practiced by Telefol women in central New Guinea and by male weaverbirds. He emphasizes not only how weaving engages bodily movement and perception with properties of materials and characteristics of the surroundings in an interactive system but also how important practice in these interactions is in developing the skill. The weaving of string bags by the women and of nests by the birds both involve collection and preparation of appropriate materials and complicated knot making, none of which can be achieved without active engagement with the materials over time. Just as children and young birds babble sounds as a prelude to speaking and singing, Telefol girls and young male weaverbirds play with fibers to develop their facility with them. Weaverbird nests are attached to branches by "a variety of stitches and fastenings" that involve a tricky operation of "threading the strip [the bird] is holding under another, transverse one so that it then be passed over the next" *(Perception* 358–59). Ingold comments: "Mastering this operation calls for a good deal of practice. From an early age, weaverbirds spend much of their time manipulating all kinds of objects with their beaks, and seem to have a particular interest in poking and pulling pieces of grass leaves and similar materials through holes. . . . Experiments showed that birds deprived of opportunities to practise and denied access to suitable materials are subsequently unable to build adequate nests, or even to build at all" (359).

Similarly, when a Telefol girl made a hopeless mess in trying to complete a string bag her mother had started, her mother told her, "You must practice to get the proper feel of looping" (356). Ingold, commenting too on attempts he and his colleagues made to learn complicated knots by following written instructions and diagrams, concludes: "It seems, then, that progress from clumsiness to dexterity in the craft of [weaving] is brought about not by way of an internalization of rules and representations, but through the gradual attunement of movement and perception" (357). Understanding skill as an interactive achievement of organisms and their environments rather than as a flash of genius—as Gregory Bateson so famously argued—emphasizes the importance of playing around with stuff (pieces of wire or grass, string, words, cell phones, computer programs) in any kind of production or invention.[4]

## Writing Is Response

Writing engages writers in a complex system that is structured by their responses to one another and to the environments created by those responses. Stephen Jay Gould called this process cultural evolution, remarking on how it proceeds so much more quickly than biological evolution, but I am arguing that both cultural and biological evolution are kinds of co-evolution, or what Maturana and Varela call structural coupling, a process through which beings whose interactions are recurrent and stable undergo "mutual congruent structural changes" (75). The examples of simple tools and languages already demonstrate how animals, including humans, change themselves and their environments through coordinated action. But with humans both the interactions and the changes are more extended and elaborate.

Bruno Latour, who argues memorably in his book *We Have Never Been Modern* that we have never escaped nature but have continually recruited it into hybrids, complicated nets of meaning and action, defines human co-evolution as a process of delegation, a "transcendence that lacks a contrary" because nothing is left behind: "The utterance, or the delegation, or the sending of a message or a messenger, makes it possible to remain in presence—that is, to exist. When we abandon the [assumptions of modernity], we do not fall upon someone or something, we do not land on an essence, but on a process, on a movement, a passage—literally a pass, in the sense of this term as used in ball games. . . . The world of meaning and the world of being are one and the same world, that of translation, substitution, delegation, passing" (129). Much as members of the New London Group see the Designer as one who refashions Available Designs into the Redesigned, Latour sees the human as "a weaver of morphisms," the mediator between subject and object, continually rearticulating those extended nets of meaning and action (137).

In concurrence with Latour, Andy Clark sees being human as a process of transformation, and especially self-transformation that is enabled by humans' "natural proclivity" for elaborating cognitive ecologies: "It is our natural proclivity for tool-based extension, and profound and repeated self-transformation, that explains how we humans can be *so very special* while at the same time being not so very different, biologically speaking, from the other animals with whom we share both the planet and most of our genes. What makes us distinctively human is our capacity to continually restructure and rebuild our own mental circuitry, courtesy of an empowering web of culture, education, technology, and artifacts" (10). Leroi-Gourhan also identified a natural proclivity for tool-based extension with what makes humans special, observing that human tools include symbols: "Humans, though they started out with the same formula as primates, can make tools as well as symbols, both of which derive from the same process or, rather, draw upon the same basic equipment in the brain. This leads us to conclude, not only that language is as characteristic of humans as are tools, but also that both are the expression of the same intrinsically human property" (113).

Jacques Derrida applauds Leroi-Gourhan for refusing to trace the origin of the human only to symbolic language, but, along with Latour and Clark, Derrida sees this "intrinsically human property" more as a process, describing it as "a stage or articulation in the history of life" (84), what he calls "différance." Derrida explains, "Instead of having recourse to the concepts that habitually serve to distinguish man from other living beings (instinct and intelligence, absence or presence of speech, of society, of economy, etc. etc.), the notion of *program* is invoked" (84), a notion that links the biological (genetic programs) with the technological (electronic programs). The emergence of the written sign, like the emergence of the tool, marks the emergence of "a 'liberation of memory,' . . . an exteriorization always already begun but always larger than the trace which, beginning from the elementary programs of so-called 'instinctive' behavior up to the constitution of electronic card-indexes and reading machines, enlarges différance and the possibility of putting in reserve" (84). Writing creates distinctions through a process of exteriorization and reification, liberating memory by turning ideas and actions into objects that can be passed to others, as Latour says, to be reinterpreted and rearticulated into new practices.

But, again, we do not simply use writing as a system extrinsic to our being to pass on our ideas and cultural practices. It is a system we are involved in, that we create in our living and that re-creates us in an ongoing way. Words and tools liberate memory not through being a record of past thoughts but by providing what Clark calls problem-solving artifacts or cognitive shortcuts that "effectively transform complex problems into ones that the biological brain is better equipped to solve" (77).[5] By now, it should not be surprising to realize that animals other than humans have these shortcuts and can be taught some of our shortcuts too. Clark describes a study that illustrates how abstract thinking is enabled by manipulating symbols. Chimpanzees were taught to associate a particular shape (a circle, for instance) with any pair of identical objects (two roosters) and a different shape (a triangle, for instance) with any pair of different objects (a rooster and pencils). They could then solve the more abstract problem of telling whether two pairs of paired objects were the same or different (see fig. 1.2): two pairs in which one pair contains identical objects and the other pair contains different objects are represented by different shapes (a circle and a triangle, respectively), and thus the pairs can be seen to be different, whereas two pairs in which each object in the pair is different from the other are each represented by the same shape (two triangles), and thus they can be seen to be the same. Just as the shapes make a concept (sameness, difference) into an object that can be manipulated and reinterpreted, words and tools enable us to play around with "stuff" and create new patterns, and then to use those new patterns to create others in levels of increasing complexity.

Clark argues that humans are "natural born cyborgs": "One large jump or discontinuity in human cognitive evolution seems to involve the distinctive way human brains repeatedly create and exploit various species of cognitive technology so as to expand and reshape the space of human reason. We—more than any other creature on the planet—deploy nonbiological elements (instruments, media, notation) to *complement* our basic biological modes of processing, creating extended cognitive systems whose computational and problem-solving profiles are quite different from those of the naked brain" (78). Clark's idea of extended cognitive systems is inspired by Edwin Hutchins's oft-cited study of navigation as an expert system in which "a good deal of the expertise in the system is in the artifacts (both the external implements and the internal strategies). . . . The system of person-in-interaction-with-technology exhibits expertise" (155), which, in turn, derives from Bateson's idea of "a flexible organism-in-its-environment" (451).

## Teaching with Technology

There are many implications for teaching writing, and especially for teaching writing with technology, of this vision of the human as a natural-born cyborg, and if you are attuned to the possibilities, you have undoubtedly already come up with many ideas about how to apply it in your own teaching. Thus I will not venture to compile a list of strategies, but I would like to briefly suggest how understanding writing in this way might alter how we think about teaching with technology. One important difference lies in how we encourage students to approach the rhetorical situation. Writers are never separate from the rhetorical

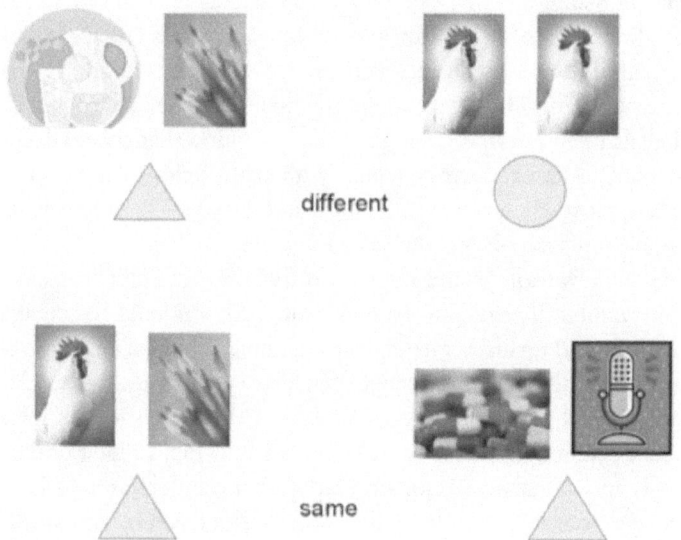

Fig. 1.2 Cognitive shortcuts

situation in which they write. They do not study the situation as something apart from them and then create in a vacuum a text that will change the situation; instead, they fully engage in the situation and respond to it. Anne Wysocki has argued that because a design approach to creating communications "has been tied to the development of useful (instead of readable) objects, it tends to foster a more concrete and bodily sense of audience, purpose, and context," and because designers tend to experiment to find what works, "by exploring and testing possibilities, they are more likely to develop what fits" (69). Understanding writing as embodied, as biological and technological as well as social and cultural, means taking a design approach to creating texts and encouraging students to do so.

Another difference is that, as Elizabeth Ellsworth and Jean Lave and Etienne Wenger have pointed out, teaching a skill is not simply a matter of detailing rules, procedures, and strategies. The direct transmission model of teaching remains influential in writing pedagogy and can lead teachers to overvalue "systematic, analytic, and conscious understanding" and undervalue practice (New London Group 35).[6] In discussing the challenges of teaching multimodal composition, Cynthia Selfe comments that in contrast to students assigned alphabetic essays who can rely on a robust understanding of written English acquired through immersion and direct instruction, students assigned audio and visual essays, "although they have been immersed in media-rich environments . . . may not have had any *direct* instruction in the genres of multimodal composing or the compositional elements that make up such genres" (17). Although her emphasis here on direct instruction suggests that the resources for teachers offered in this edited collection might consist of definitions of multimodal genres and their elements, the chapters that follow instead recommend experimentation and open-ended and flexible assignments. Understanding the acquisition of writing skills as a matter of gradual attunement of movement and perception that comes dominantly through practice, a lot of playing around with stuff, helps us remember that what students lack most when faced with audio and visual essay assignments is any experience—and practice—in producing such texts.

In closing, here are a few more examples, drawn from Stuart Selber's discussion of functional computer literacy, of how encouraging students to engage with rhetorical situations and technologies leads to the emergence of capabilities that are not individual but rather the property of cognitive ecologies—organism-persons interacting in richly structured environments. As with the examples drawn from Wysocki's and Selfe's work, Selber's discussion evinces a tacit understanding of writing as an emergent, embodied response, an understanding I am simply trying to make more recognizable. Selber argues that teaching functional computer literacy involves enabling students to deal with educational goals, social conventions, specialized discourses, management activities, and technological impasses, and he describes some activities he finds most useful to "help programs and teachers develop their own" activities (475). The activities he suggests are described as explicit procedures and strategies, but they can be framed more productively to engage students in the systems of working with computers.

For example, to help students understand social conventions of computer use, he asks them to use a taxonomy of types of unacceptable behavior to analyze the conventions of a newsgroup they are interested in. He cautions, however, that the assignment can lead students to overgeneralizations about the norms for behavior in all newsgroups, norms that have actually proved to be highly localized and "still in a somewhat embryonic state" (483). If, instead of asking students to objectively analyze a system from the outside, we ask them to think about their own engagement in an interobjective system, students can understand how norms emerge from users' coordinating their behaviors to achieve benefits. Thus we might ask students what they like about a particular newsgroup they participate in and what behaviors enhance or detract from their enjoyment or benefit. Asking students about what kinds of behaviors draw complaints from other users (and, more important, why) and how these behaviors conflict with the purposes of the newsgroup focuses their attention on the purpose or value of a particular interaction in a particular context and how certain behaviors create ongoing consensual domains that offer specific benefits to users. How computer users tend to interact with computers in ways that benefit them rather than trying to learn how to use them first, like the crows Betty and Abel who get the meat not through first acquiring the technical skills needed to use hooks but through interacting in beneficial ways with their environment, is also demonstrated by Selber's students' reluctance to use email filters to manage their coursework. As he concedes, setting up filters makes sense only in a situation where there is a reason to commit to long-term managerial structures, and in most courses lasting only ten to fifteen weeks deleting or manually sorting messages is easier (492).

Selber suggests that dealing with technological impasses is like locating the exigency in a rhetorical situation: "the key is to situate technological impasses in a broader context so that their characteristics can be organized and understood" (495–96). He offers a "relatively simple heuristic" that involves phrasing the impasse as a qualitative question, locating the question in a matrix of five categories of computer-user concerns, and matching the categories to types of assistance that will enable students to resolve their problem. Apart from not being a particularly simple task, this procedure construes the technology as a tool separate from its user, not as something that arises out of an engagement in a rhetorical situation. Students "who think that the only way to turn the grammar checker off is to stop writing 'ungrammatical' sentences" have not so much misunderstood themselves as "the causal root of technological impasses" (496) as mistaken a rhetorical problem for a technological one. Asking themselves what rhetorical exigency led to the emergence of grammar checkers (as well as whether grammar checkers usefully respond to that exigency) will lead students to a variety of ways to resolve their impatience with squiggly lines in their texts: deciding that "correct" grammar does not matter in this rhetorical situation (and therefore that turning off the grammar checker makes more sense than trying to change their texts), or asking their teacher whether and why teachers require the use of grammar checkers, for example.

This is the mistake Horkheimer and Adorno identified as the disaster of Enlightenment thinking, which they trace to the shift in classical times from seeing language and technology as living forces in the world to seeing them as instruments alien to the natural world and alienating to the human psyche. Heidegger, too, sought to liberate Western culture from technological "enframing" that converts every thing and every being into "standing reserve." Understanding writing as a complex system in which human interactions elaborate cognitive ecologies allows us to understand words and tools as Ingold suggests we should, as mediating our active engagement with our environment rather than asserting our control over it. Far from alienating us from the world and our own natures, words and tools connect us inextricably to others and to our environment and make us what we are, the animal who writes.

## Notes

1. For details of the history of complexity theory, see Capra, Harrington, Hayles, Taylor, and Waldrop.

2. See also Engeström and Miettinen, who in their argument for the relevance of activity theory similarly observe connections between distributed cognition, situated

learning, actor networks, and the rejection of "monocausal explanation" in the new sociology of science (8–9).

3. Margaret Syverson, Jay Lemke, and most recently Byron Hawk are notable exceptions.

4. As Johnson-Eilola writes of asking his daughter Carolyn how she figured out the rules of a new computer game she had downloaded, she answered: "You just . . . play" (3).

5. The methods Selber cites of converting syntactic technical knowledge into semantic (and more easily remembered) knowledge are other examples of such cognitive shortcuts.

6. The New London Group points out that "Overt Instruction does not imply direct transmission, drills, and rote memorization . . . Rather it includes all those active interventions on the part of the teacher and other experts that scaffold learning activities" (33), and they emphasize the fact that "Situated Practice does not necessarily lead to conscious control and awareness of what one knows and does" (32).

## Works Cited

Barthes, Roland. "The Death of the Author." In *Image, Music, Text*. Translated by Stephen Heath. New York: Hill and Wang, 1977.

Bateson, Gregory. *Steps to an Ecology of Mind*. New York: Ballantine, 1972.

Bazerman, Charles, and David R. Russell. "Introduction." In *Writing Selves/Writing Societies*, edited by Charles Bazerman and David R. Russell, 1–6. 2003. Accessed 26 January 2009, http://wac.colostate.edu/books/selves_societies/.

Capra, Fritjof. *The Web of Life: A New Scientific Understanding of Living Systems*. New York: Anchor Books, 1996.

Clark, Andy. *Natural-Born Cyborgs: Minds, Technologies, and the Future of Human Intelligence*. Oxford: Oxford University Press, 2003.

Csányi, Vilmos. *If Dogs Could Talk: Exploring the Canine Mind*. Translated by Richard E. Quandt. New York: Farrar, Straus, and Giroux, 2000.

Davydov, Vassily V. "The Content and Unsolved Problems of Activity Theory." In *Perspectives on Activity Theory*, edited by Yrjö Engeström, Reijo Miettinen, and Raija-Leena Punamäki, 39–52. Cambridge: Cambridge University Press, 1999.

Derrida, Jacques. *Of Grammatology*. Translated by Gayatri Chakravorty Spivak. Baltimore: Johns Hopkins University Press, 1974.

Ellsworth, Elizabeth. *Teaching Positions: Difference, Pedagogy, and the Power of Address*. New York: Teachers College Press, 1997.

Engeström, Yrjö, and Reijo Miettinen. "Introduction." In *Perspectives on Activity Theory*, edited by Yrjö Engeström, Reijo Miettinen, and Raija-Leena Punamäki, 1–16. Cambridge: Cambridge University Press, 1999.

Friend, Tim. "Crows Exceed Expected Intelligence Levels." *USA Today*, 9 August 2002. Accessed 2 February 2005, http://www.usatoday.com/news/science/2002-08-08-smart-crows_x.htm.

Gleick, James. *Chaos: Making a New Science*. New York: Viking, 1987.

Gould, Stephen Jay. *The Mismeasure of Man*. New York: Norton, 1981.

Griffin, Donald R. *Animal Minds*. Chicago: University of Chicago Press, 1994.

Hansen, Mark B. N. *Bodies in Code: Interfaces with Digital Media.* New York: Routledge, 2006.
Harrington, Anne. *Reenchanted Science: Holism in German Culture from Wilhelm II to Hitler.* Princeton, N.J.: Princeton University Press, 1996.
Hauser, Marc D. *Wild Minds: What Animals Really Think.* New York: Henry Holt, 2000.
Hawk, Byron. *A Counter-History of Composition: Toward Methodologies of Complexity.* Pittsburgh: University of Pittsburgh Press, 2007.
Hayles, N. Katherine. *How We Became Posthuman: Virtual Bodies in Cybernetics, Literature, and Informatics.* Chicago: University of Chicago Press, 1999.
Hoagland, Edward. "Small Silences: Listening for the Lessons of Nature." *Harper's Magazine* (July 2004): 45–58.
Heidegger, Martin. "The Question Concerning Technology." In *Basic Writings*, edited by David Farrell Krell, 283–317. New York: Harper & Row, 1977.
Horkheimer, Max, and Theodor W. Adorno. *Dialectic of Enlightenment.* Translated by John Cumming. New York: Continuum, 1989.
Hutchins, Edwin. *Cognition in the Wild.* Cambridge, Mass.: MIT Press, 1995.
Ingold, Tim. *The Perception of the Environment: Essays in Livelihood, Dwelling, and Skill.* London: Routledge, 2000.
———. "Tool-Use, Sociality and Intelligence." In *Tools, Language and Cognition in Human Evolution*, edited by Kathleen Gibson and Tim Ingold, 429–45. Cambridge: Cambridge University Press, 1993.
Johnson, Steven. "Tool for Thought." *New York Times*, 30 January 2005. Accessed 2 February 2005, http://www.nytimes.com/2005/01/30/books/review/30JOHNSON.html?emc=eta1.
Johnson-Eilola, Johndan. *Datacloud: Toward a New Theory of Online Work.* Cresskill, N.J.: Hampton Press, 2005.
Latour, Bruno. *We Have Never Been Modern.* Translated by Catherine Porter. Cambridge, Mass.: Harvard University Press, 1993.
Lave, Jean, and Etienne Wenger. *Situated Learning: Legitimate Peripheral Participation.* Cambridge: Cambridge University Press, 1991.
Lemke, Jay. *Textual Politics: Discourse and Social Dynamics.* London: Taylor & Francis, 1995.
Leroi-Gourhan, André. *Gesture and Speech.* Translated by A. B. Berger. Cambridge, Mass.: MIT Press, 1993.
Mandelbrot, Benoît. *The Fractal Geometry of Nature.* New York: W. H. Freeman, 1982.
Maturana, Humberto. "Biology of Language: The Epistemology of Reality." In *Psychology and Biology of Language and Thought: Essays in Honor of Eric Lenneberg*, edited by George A. Miller and Elizabeth Lenneberg, 27–63. New York: Academic Press, 1978.
———. "The Nature of the Laws of Nature." *Systems Research and Behavioral Science* 17 (2000): 459–68.
Maturana, Humberto, and Francisco Varela. *The Tree of Knowledge: The Biological Roots of Human Understanding.* Rev. ed. Boston: Shambhala, 1998.
Merleau-Ponty, Maurice. *Phenomenology of Perception.* Translated by Colin Smith. London: Routledge, 1962.

New London Group. "A Pedagogy of Multiliteracies: Designing Social Futures." In *Multiliteracies: Literacy Learning and the Design of Social Futures*, edited by Bill Cope and Mary Kalantzis, 9–37. London: Routledge, 2000.

Prigogine, Ilya, and Isabelle Stengers. *Order Out of Chaos: Man's New Dialogue with Nature*. New York: Bantam, 1984.

Prior, Paul, and Jody Shipka. "Chronotopic Lamination: Tracing the Contours of Literate Activity." In *Writing Selves/Writing Societies*, edited by Charles Bazerman and David R. Russell, 180–238. 2003. Accessed 26 January 2009, http://wac.colostate.edu/books/selves_societies/.

Russell, David. "Activity Theory and Its Implications for Writing Instruction." In *Reconceiving Writing, Rethinking Writing Instruction*, edited by Joseph Petraglia, 51–77. Mahwah, N.J.: LEA, 1995.

Selber, Stuart A. "Reimagining the Functional Side of Computer Literacy." *College Composition and Communication* 55 (2004): 470–503.

Selfe, Cynthia L., ed. *Multimodal Composition: Resources for Teachers*. Cresskill, N.J.: Hampton Press, 2007.

Syverson, Margaret. *The Wealth of Reality: An Ecology of Composition*. Carbondale: Southern Illinois University Press, 1999.

Taylor, Mark C. *The Moment of Complexity: Emerging Network Culture*. Chicago: University of Chicago Press, 2001.

Uexküll, Jakob von. "A Stroll through the Worlds of Animals and Men: A Picture Book of Invisible Worlds." Translated by Claire H. Schiller. *Semiotica* 89.4 (1992): 319–91. First published 1957 in *Instinctive Behavior* (Claire H. Schiller, ed.) by International University Press.

Vitanza, Victor J. "Abandoned to Writing: Notes toward Several Provocations." *Enculturation*, Fall 2003. Accessed 26 January 2009, http://enculturation.gmu.edu/5_1/vitanza.html.

Waldrop, M. Mitchell. *Complexity: The Emerging Science at the Edge of Order and Chaos*. New York: Simon & Schuster, 1992.

Winograd, Terry, and Fernando Flores. *Understanding Computers and Cognition: A New Foundation for Design*. Norwood, N.J.: Ablex, 1986.

Wittgenstein, Ludwig. *Philosophical Investigations*. 3rd ed. Translated by G. E. M. Anscombe. New York: Macmillan, 1968.

Wray, Alison, ed. *The Transition to Language: Studies in the Evolution of Language*. Oxford: Oxford University Press, 2002.

Wysocki, Anne Frances. "Using Design Approaches to Help Students Develop Engaging and Effective Materials that Teach Scientific and Technical Concepts." In *Resources in Technical Communication: Outcomes and Approaches*, edited by Cynthia L. Selfe, 63–89. Amityville, N.Y.: Baywood, 2007.

# Among Texts

Johndan Johnson-Eilola

This will to truth, like the other systems of exclusion, relies on institutional support: it is both reinforced and accompanied by whole strata of practices such as pedagogy—naturally—the book-system, publishing, libraries such as the learned societies in the past, and laboratories today. But it is probably even more profoundly accompanied by the manner in which knowledge is employed in a society, the way in which it is exploited, divided and, in some ways, attributed. (Foucault 219)

Ambient informatics is a state in which information is freely available at the point in space and time someone requires it, generally to support a specific decision. Maybe it's easiest to describe it as the information detached from the Web's creaky armature of pages, sites, feeds, and browsers, and set free instead in the wider world to be accessed how and where you want it. (Greenfield 24–25)

Contrary to a deeply rooted belief, the book is not an image of the world. It forms a rhizome with the world, there is an aparallel evolution of the book and the world; the book assures the deterritorialization of the world, but the world effects a reterritorialization of the book, which in turn deterritorializes itself in the world (if it is capable, if it can). (Deleuze and Guattari 11)

This essay examines *reading texts* in the postmodern sense that any object, collection of objects, or contexts can be "read" by tracing and retracing the slipping, contradictory network of connections, disconnections, presences, absences, and assemblages that occupy problematic spaces (both conceptually and physically) at the margins (transgressive like a pun) of what it means to write and read texts, within texts, and among texts.

This tangled space has many ways into it; one is shown in figure 2.1.

If postmodernism assumes that any object can be treated as pages of a book, Geoffery Rockwell's image inverts that question: "What if we treat pages as matter?" Rockwell's images (fig. 2.1 is part of a series)—grainy, monochromatic, and simultaneously kinetic and ponderous—show books variously cut apart, drilled, sanded, and otherwise mutilated. These are not the ways most of us are comfortable using books. Books are things to be consulted for their wisdom. The cultural history of books weaves strands of literal religious reverence and scholarship, beginning with monastic scholarship, accelerating with the Gutenberg Bible, and continuing into the construction of libraries as contemplative spaces (fig. 2.2).

But if the library at the Strahav Monastery (Morroia) suggests the still-current image of books as religious, contemplative, quiet objects—the image *works* for contemporary users because it plays on still-functional forces that construct all libraries as religious spaces—other emerging cultural forces push against that particular articulation.

Although the history of books may have been partially structured as an appreciation of sacred objects, that history also includes other social and technical forces: book as functional object, book as conversation, book as commodity, book as distribution of knowledge. These shifts may seem to us very recent, but the seeds of these shifts have been present for quite some time—arguably for the whole history of books. The meaning of texts has long been contentious—not merely the literal meaning of books as objects, but the social meaning of literacy itself. The social upheaval enabled by the mass production of the Christian Bible, one of the most discussed examples in our culture, involved (among many other things) seismic shifts in what it meant to be a scholar. Access to sacred texts afforded access to thinking with sacred texts: no longer simply told what rare texts meant by religious authorities, the growing literate population could now examine texts in detail and arrive, potentially, at conclusions that contradicted the official positions of churches.

Although my thumbnail sketch of the cultural and technological history of religious thought is almost laughable in its broad strokes and omissions, I think these rough outlines are enough to help us understand how we have come to our somewhat contradictory impulses in how we understand and use texts of nearly any type: simultaneously revered and used, worshiped and put to work.

Agostino Ramelli's "book-wheel" (fig. 2.3) was sketched by an Italian engineer who worked for King Henry III (Goodall). One source gave this amusing description of the device: "This is a beautiful and ingenious machine, very useful and convenient for anyone who takes pleasure in study, especially those who are indisposed by gout. For with this machine a man can see and turn through a

Among Texts  35

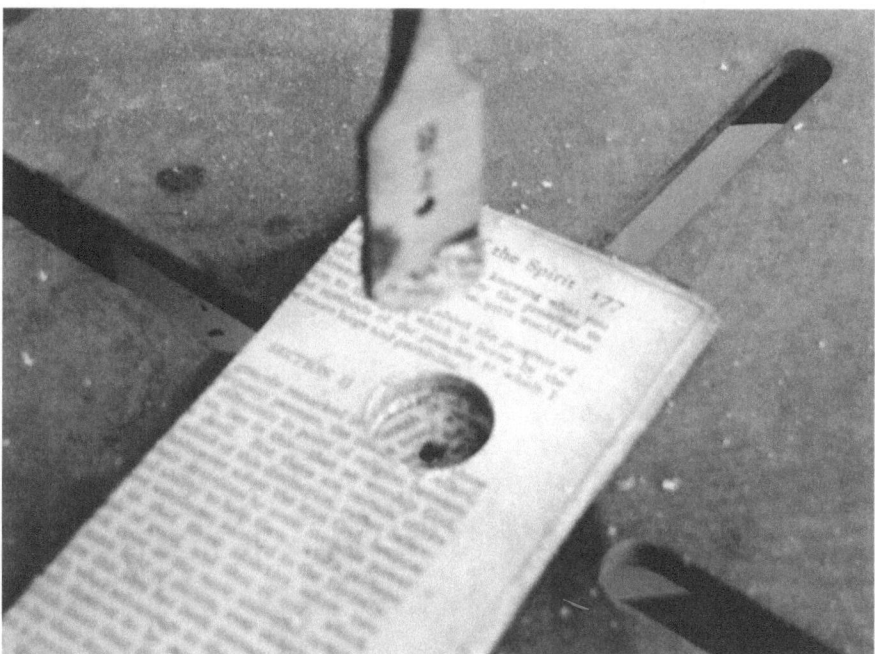

Fig. 2.1 Image from "Text in a Machine" series. Photograph courtesy of Geoffrey Rockwell

Fig. 2.2 Strahav Monastery library. Courtesy of Creative Commons Attribution 2.0 Generic (Morroia)

large number of books without moving from one spot. Moreover, it has another fine convenience in that it occupies very little space in the place where it is set, as anyone of intelligence can clearly see from the drawing" (qtd. in Basbanes 291).

The book-wheel, opposed to the cathedral library, highlights the ways scholars have long struggled to force books to move in more fluid ways in support

Fig. 2.3 Ramelli's book-wheel

of intellectual work. Books are wondrous things, but ultimately they do not do what we want them to do.

## Texts, Work, and Agency

The book-wheel reminds us that we have long felt a contradictory need to not simply *appreciate* texts but also put them to work: becoming literate is a complicated, often contentious activity. The terrain has obviously shifted in the last several decades, with examinations of dramatic and mundane changes in what it means to be literate in an extensively mediated, networked culture.

These examinations suggest how complicated our relationships to texts have become recently, but I would like to push this question slightly further into the technical sphere to ask, What happens when our texts become *actively* social? It is one thing to read a multimedia, networked text with leaky boundaries; it is another to read a text that itself has intentions and agency that do not so much leak as roll like a river and babble like a brook. I am not suggesting that writers and readers have no agency, but only saying that the whole issue gains an extra level of intractable complexity when texts themselves are not merely *out there*, as objects, but also in motion, gathering other texts around them, responding to their environments in ways both simple and complex, making connections that their authors or readers are participants in, rather than simple agents of—intertextuality with teeth.

We can be forgiven for not having adjusted or even remarked much on the idea of texts as agents. The signs we thought would mark this event are based on other, flashier technologies: the panoptic, sterile eye of HAL in *2001: A Space Odyssey*, animated buckets and brooms in *The Sorcerer's Apprentice*, the golem. What these agents all share is the spark of language in one form or another: Dr. Chandra programming HAL, Mickey's bumbling the recitation of spells from the sorcerer's book, Rabbi Löw inscribing the word of God on a tablet and placing it under the golem's tongue: "And the Lord God formed man of the dust of the ground, and breathed into his nostrils the breath of life; and man became a living soul" (Gen. 2.7, King James [Authorized] Version).

If texts have power, we are often reminded that texts can also have too much power: autonomous texts slip human control and eventually turn on their masters. We are already partway there: databases, computer forms, video games, and many other things we regard provisionally as texts have already begun to gain agency. Such semiautonomous texts are still relatively isolated and only slightly active: they respond to our touch, they provide spaces where we talk to one another (and to the texts), they record some small portion of our actions. But what will it mean when these texts become more aware of their surroundings, of the other texts that they bump up against?

Here I take a substantial detour into a sketchy history of one class of manufactured objects, documented by writers such as Bruce Sterling ("Dumbing Down";

*Shaping Things*) and Adam Greenfield (*Everyware*). This history focuses on the textual nature of objects, particularly certain increases in agency as some mass-produced objects become more aware of and active within their surroundings.

## A Brief History of Spimes: Artifact, Product, Gizmo, Spime

To map better the trajectory and shape of the changing nature of texts, I draw heavily on Bruce Sterling's account of the rise of one type of communicative object, the "spime." Spimes are simply objects that are aware of their own contexts and communicate about those contexts, usually using relatively cheap, wireless, networked sensors.

Sterling's examples include a spimelike tennis shoe that theoretically could track simple activities like a user's gait, impact patterns, wear patterns, and other activities over time. The spime-enabled shoe could then, using cheap wireless chips, transmit information about itself to networked computers, mapping gathered data over time against variables such as improvements or declines in ability, changes in terrain or exercise regimen, health, and other factors. Additionally a spime-enabled shoe could communicate information that allows manufacturers to gather usage data that could assist in redesigning shoes, suggesting to users specific changes in exercise patterns, and so on.

This example is intentionally mundane but also very powerful: tracking small changes among a large number of variables can be extremely helpful in assessing and improving performance. The key to spimes is exactly their applicability in mundane situations: spimes are both very cheap (on the order of a few dollars or even far less) and very small (at the sub-millimeter level). Because they generally communicate wirelessly, using very low (or no) built-in power, they can be inserted into an enormous range of objects, even subdermally into adventurous users to allow them to unlock their car and apartment doors with a wave of their microchip-enhanced hands (Graafstra).

The cultural history of technology is rife with examples of seemingly minor objects inserted into existing objects and situations in transformative ways: the transition from manuscript to print (and the corresponding increase, through economies of scale, in print literacy), the invention of the transistor, the spread of accurate clocks, the rapid adoption of HTML and HTTP protocols (which many experts discounted as clumsy, limited versions of already existing network communication protocols). Transformative technologies, of course, can transform situations or cultures through massive, highly visible methods or small, under-the-radar ways and, in some cases, methods both large and small, slow and fast. Skyscrapers and mass transit both have origins in much more mundane activities (building small dwellings; herding cattle) that, over time, led to much larger infrastructural developments. Spimes, at this point, are still at the thin and dispersed edge of infrastructural change and therefore work (socially and technically) when they are inserted into already existing technologies and patterns of use.

Sterling's brief but sweeping history of spimes illustrates how spimes can act on existing technologies by accelerating tendencies built into various objects over long periods of time. Spimes offer possibilities because they redirect already existing tendential forces. In recounting his history, I view Sterling's account (which offers an enormous range of technologies, including the tennis shoe described earlier as well as wine bottles, shampoo containers, and more) through the lens of the development of texts as technologies, beginning with the manuscript and moving forward through networked texts.[1] Table 1 provides a quick summary of key points.

Text as Artifact

Texts were initially handmade objects: markings on cave walls, clay counters, and illustrated manuscripts (fig. 2.4). They were made individually, often laboriously, and used in the same way: carefully, slowly, and for very specialized situations.

At this stage, texts have very isolated (although still possibly very important) uses in specific situations. Using an artifactual text is a relatively special event. Texts do not communicate with each other except by human intervention (an aspect that will continue through many stages of the evolution of texts): a person reads one text and in some cases copies some portion of it into another text. In the case of duplicating manuscripts, scribes copy the full manuscript and attempt, with various degrees of success, to avoid introducing errors or other unintentional changes to the original. In one way of thinking, every artifactual text is an isolated, original, static object. Using artifactual texts—literacy—represents a profound shift in work flow and thought process, both individually and socially. Compared to later textual technologies (products, gizmos, spimes),

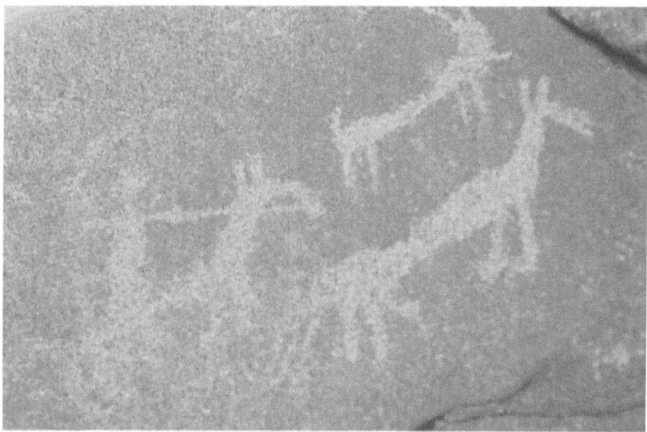

Fig. 2.4 Text as artifact: petroglyphs in Chilas, Pakistan. Courtesy of Creative Commons Attribution 2.0 Generic (Morroia)

Table 1

# Four of Sterling's Technology Types

| Stage | General Characteristics | Examples | In Text Format |
|---|---|---|---|
| Artifact | Hand produced, used by hand using rules of thumb or local conventions/crafts. Very static, one-to-one usage. | Cave drawings, hand-milled screws, other crafted objects. | Illuminated manuscripts, handwritten letters. |
| Product | Mass-produced, widely distributed. Very static, one-to-one usage, but habits of use are systematized (in informal or formal education). | Automobiles, shoes, television sets, point-and-shoot cameras. | Trade market books, textbooks, logos on clothing. |
| Gizmo | Mass-produced but to niche markets. Complex workings. Users rely on both formal documentation and experimentation (extensive and ongoing: the bleeding edge). | Linux-enabled smartphones, Tenori-on musical instruments, Linux computers. (Actual examples change over time as more people adopt a technology and the "edges" are worn off through refinement and training.) | Tinderbox, Web pages (to some extent), Delicious Library book-tracking system, Google Analytics. |
| Spime | Mass-produced and widely distributed. Often operate automatically with occasional intervention/reference by manufacturer and user. Also allow gizmo-like experimentation and change. | Nike shoes with embedded sensors, RFID-enabled badges. (Not currently a very well-developed stage.) | Texts tracking their own readings, citations, quotations; texts gathering emerging new texts on related topics; texts noticing how they are used in relation to other texts (virtual or physical). |

Fig. 2.5 Masses of mass-produced books

however, artifactual texts seem very clumsy. Everything is relative, though, and artifactual texts still retain value, even as later types are adopted: most knowledge workers still keep handwritten notes and journals and make marginal notes (modern palimpsests) on later types of texts.

Text as Product (Rise of Mass Production)

As texts began to be used more widely, the economies of scale and time supported by institutions such as monasteries brought about situations in which coordinated, repeatable communication was necessary: the hierarchal organization of religious institutions, with outposts scattered in increasingly wide ranges, created a need for mass duplication of texts. (Talking about agency in a one-way fashion in technology development presents serious problems: it is a moot point whether the spread of religious institutions *created* a need for mass-produced texts or whether the tentative adoption of mass-produced texts *created* the possibility of wider-spreading religious institutions.)

The point here is that transformative technologies can emerge relatively slowly by their gradual adoption and their integration into existing patterns of work and communication. At this stage, texts are still slightly awkward objects in terms of work flow: less difficult for knowledge workers to use than artifactual texts but of limited help in tracking, organizing, sorting, and storing product texts (fig. 2.5). Much of the work life of a knowledge worker involves the administrative-clerical work of simply moving things around. There is some inherent value, undoubtedly, to the physical and spatial movement of texts:

Fig. 2.6 Workspaces: (left to right) Johndan Johnson-Eilola, Dennis Jerz, Charlie Lowe. Photographs courtesy of Dennis Jerz and Charlie Lowe

juxtaposition, chance, memory, and creativity often emerge in the midst of this seemingly mundane work. But we might question whether or not we could do less of it, or make it easier to do, in order to focus on other things. I am not sure how much of my own appreciation of printed books is simply nostalgia—certainly not all of it, but just as certainly *some* of it.

Due to the benefits of widespread literacy education for many, the immediate use of books is relatively straightforward: one sits and reads.[2] But for symbolic-analytic workers who work with texts, the activity of *using* a book involves not simply reading but also collecting, arranging, synthesizing, and filtering all of the texts one reads. And as anyone who works with texts for a living knows, the full effort of actually *using* a text in a productive way frequently takes place in immensely complex and inefficient (if much-cherished) environments, such as the ones shown in figure 2.6.

For many who work with texts, isolated and physical texts are still one of our primary environments. This is true even for many people who spend much of their productive time online, given (a) the sheer amount of important information that is still primarily available in print, and (b) the relatively slow pace of change in academia, which still values isolated printed texts.

Text as Gizmo (Power Users, Tinkering)

Sterling's gizmo category includes, as the name suggests, objects that are remarkable for their cunning and complex nature. Operating them usually requires some effort and learning, and not a little experimentation, on the part of users. Gizmos malfunction and they break down. They are powerful and fun, but also frustrating. They are often thought of as "cutting edge"; they are "sharp" both in the sense of being stylish and of being somewhat dangerous. As the sharp edges are worn away (both through technological refinements and through broader, easier cultural adaptations), they tend to become products but may also become spimes. So a gizmo is, culturally speaking, an object that tells its own story about its potential future: hiding its flaws, flaunting its power, hoping for continued technological development that will allow it to achieve the fiction that it is trying to tell.

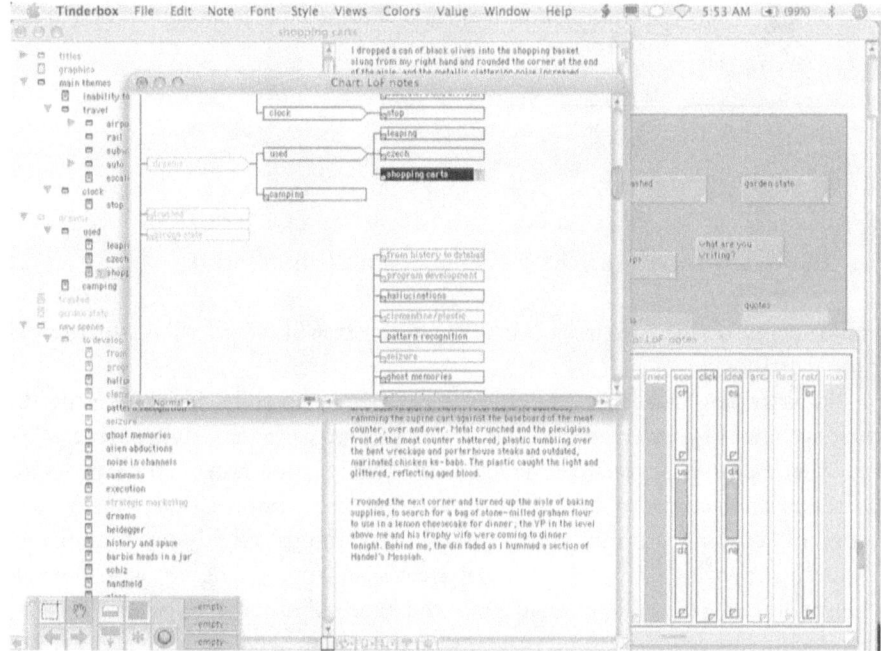

Fig. 2.7 Text as gizmo: Multiple views of one text in Tinderbox

Paradoxically, gizmos are both highly visible and commented upon but relatively rare compared to products or artifacts. Media tend to fixate on them because they are noteworthy and promise much power; their failures are not often publicly discussed, at least initially (the degree to which the failures do become public corresponds to the likelihood that the gizmo will disappear: a terminating branch in development).

Text in the gizmo format represents a dramatic departure from text as product, although we are still—after decades—coming to grips with that transition. (The pace of this shift, however, is much faster than the previous transition from artifact to product.) As gizmos, texts are highly unstable and user-alterable in ways that printed texts are not: They can be moved around, recombined, and transformed in useful and sometimes surprising ways.

But as with other types of gizmos, these texts are often awkward and difficult to use: Users are on the bleeding edge of technology, where things break down frequently and sometimes abruptly. Programs running such texts crash or perform in ways we had not anticipated. Files are overwritten, data lost, information mangled beyond recognition. But when they work—extremely cool.

Not surprisingly, users of gizmos tend to be relatively rare, given the amount of effort (and risk) entailed. Learning how to use a gizmo such as a Tinderbox text (fig. 2.7), for instance, involves a steep learning curve, with frequent reference to

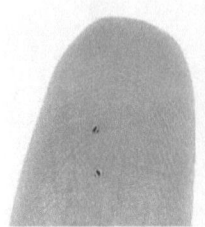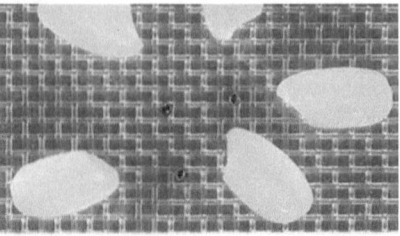

Fig. 2.8 Two views of Hitachi's μ-Chip (0.4mm x 0.4mm)

manuals and online resources (including discussion lists and wikis, themselves textual gizmos).

Summarizing Tinderbox succinctly is nearly impossible. Eastgate Systems, the developer of Tinderbox, describes the program as "a personal content assistant that helps you visualize, analyze, and share your notes." The program possesses characteristics of database, text processor, information visualization program, and brainstorming tool, among other things. Highly programmable, Tinderbox provides a free-form database for storing, arranging, and filtering chunks of media (words and images). In the hands of experienced users, Tinderbox documents can gather and rearrange information according to rules written by users. In the hands of novices, Tinderbox can be a little baffling. Tinderbox, then, is a gizmo for people who work with texts. At the same time, Tinderbox can also be an environment for developing spimes.

Text as Spime (Semiautonomous, Networked)

The transition to spimes involves a shift toward semiautonomy. Spimes communicate about themselves, gathering data about use and then sending and receiving data from a larger network. We might classify current technologies such as cell phones and computers as near spimes, because in some cases they are able to gather data about their surroundings (GPS location, motion, visuals, and so on) and then communicate that data automatically to other locations while downloading new information at the same time.

Such uses seem relatively routine at times, but the dramatic miniaturization of computer chips and network technologies now allows spime capabilities to be embedded in objects very different from what we are used to. Chips and antennae, such as Hitachi's μ-chip (fig. 2.8), which measures 0.4 millimeter across (including an embedded wireless antenna), allow the use of radio frequency identification tags (RFIDs) in a much broader range of objects. The parallel reduction of cost—simple RFIDs are already available for ten cents or less—means that everyday objects, including disposable items, are beginning to enter the spime realm.

Much of the public discussion about the development of spimes has focused on use of the technology in these ubiquitous consumer and household products.

Sterling's examples include the tennis shoe discussed earlier and bottles of wine. Such developments may have a large impact on our daily lives via small but transformative uses. But for academics and others who work in symbolic information, spimes may dramatically increase the flow of metainformation across physical and virtual realms in ways that disrupt our current work patterns. As with many technical developments (particularly in media), spimes remediate their technical ancestors: When we look at our old-fashioned print books and journals, we sometimes lament their static, immobile, dead nature. In other spheres, parallel developments have addressed the gaps between physical products and networked texts in ways that bridge each: Is it difficult to physically, manually count boxes of lightbulbs to enter them into the online product inventory database? Fine, we will add a wireless chip to each box so that it can shout out its name to the database when it enters or leaves the warehouse or the store. Product becomes partially virtual; information becomes partially physical. For texts, scanning a barcode or typing a title into the search field of Amazon or Google functions as a precursor to spimes: the object-as-information becomes both physical and virtual.

> In fact, books are already well on their way to becoming spimes, thanks mostly to Amazon.com. A book listed on that site is much more than the words between its covers. It looks, feels, and behaves like an ordinary book, yet in short order, you can find out its cost, publisher, and printer; whether other editions have been published and what they look like; what other books the author has written; what readers think of the book and what other books those readers have bought; what other publications quote the book; and so on. And, beyond Amazon.com, you can learn about the composition of the paper, how long it will last before yellowing, and what kinds of products it can become when the book is recycled. Some of this information might be contained in the pages, and some might be conjured on the Web via, say, an RFID tag—but in practice, it won't make much difference. The upshot is that the object's nature is transparent: an open book. (Sterling, "Dumbing Down")

Apart from this tenuous but still productive connection between physical object and information object, we have already started to work with self-aware, completely networked, and virtual objects in areas such as Weblog TrackBack, 'bots, and server log tracking. In a post at the Weblog run by Seth Godin (fig. 2.9), for example, readers see not only comments by other readers, but an extensive amount of information about what readers have done with Godin's post: the number of times it has been posted to Digg (a social ranking system that gathers links to interesting articles), saved to Delicious.com (a site for saving and sharing bookmarks to Web sites), reviewed at StumbleUpon (a site for sharing

Fig. 2.9 Partial listing of reader activity on a Weblog post by Seth Godin

brief reviews of Web content), and linked to other Weblogs via the TrackBack system. Notably such information is automatically gathered and updated by many current Weblog publishing systems, making the act of tracking conversations *about* a specific text a relatively automatic process. These activities suggest some of the possibilities of making physical texts more spimelike.

Despite the rapid growth of Weblogs over the last decade, the genre still covers only a tiny sliver of important (and mundane) information being produced. In addition to the invisibility of a wide range of Web-based information, tracking search engines such as Technorati cannot possibly track the massive amount of information still being published in print (let alone the immense existing print archive) or information published in unindexed databases or behind pay firewalls. Many savvy scholars have hacked together methods for filling in a few of the gaps—RSS (Really Simple Syndication) Web feeds of Google searches to follow new items added in other areas of the Web, manual searches of proprietary citation databases, occasional Amazon searches for new citations, and so on.

But as Bruce Sterling points out ("Dumbing Down"), hacking things together is a gizmo activity; spimes are what we get after the kinks are worked out and the information gathering and sharing are routine, widely used, easy, and at least semiautomatic.

### Implications of Texts as Spimes

To understand what text spimes might afford us, we need to step back and consider a couple of the key ways that academics in rhetoric and composition currently use text. The following abbreviated list suggests a few of the key areas. We take in a wide range of textual sources:

> print books and journal articles
> online journal articles and (increasingly) books
> calls for papers, submission guidelines, and so on
> database search engine query results (both public, such as Google, and private, such as those accessed via libraries)
> PowerPoint, Keynote, and other presentation formats from conferences (either ones we attend or those we scavenge from other sources)
> video and audio feeds and clips (conference podcasts, archival footage, automated search query updates)
> RSS feeds or other automated notifications

Those working in more deeply online areas can likely extend the list substantially in that realm, and those working in more traditional archival areas might include a much larger number of print resources as well as emerging online archival tools.

Along with this list, our workflow and work environment will likely also include the need to work with (by referring to, creating, and manipulating) additional textual information:

> notes about work being planned in computer files, paper notebooks, scraps of paper, whiteboards
> annotations written in printed texts (books, journals)

clippings and photocopies
digital photographs and video of primary research
working drafts of articles, chapters, presentations, in print or online
annotations written on physical drafts and in online documents

Again, this list is very sketchy. Note that to this point the lists have only touched on communications that are relatively less immediate. A similarly long and yet still incomplete list of texts emerges when we begin to consider communications that involve relatively direct collaboration:

email messages
instant messages
phone messages
feedback on plans, drafts, revisions
written comments on plans, drafts, revisions
notes about face-to-face or phone discussions in handwritten or
    online format
Twitter, Facebook, other social media updates

The point of this long, still-partial list is merely to suggest that the majority of us work extensively with an enormous range of material in diverse formats and physical locations. Together these somewhat tedious lists highlight something we seldom think about: Maybe we could make some aspects of this work a little easier and more efficient. We could determine which aspects of our practices we might want to change in light of these technologies and which we want to keep.

In implementing this effort, consider that merely collecting the sources listed is itself a monumental task (even if we rarely notice it, we devote a great deal of effort to it). But the actual *work* with this information involves symbolic-analytic work of a second order. This last, short list is the important one:

filtering
sorting
connecting
synthesizing
sharing

In other words, an extremely important part of our work involves making these disparate pieces of data aware of one another.

What spime texts might offer to this work environment and work flow is at once very simple and very powerful: an informational bridge across the spaces separating these disparate bits of information. Given the relatively low cost of spime technologies, it is well within the realm of possibility to consider a work environment in which many if not all of these pieces of information are aware of one another and able to note when one piece of information is stored or moved (physically or virtually) within the proximity of another:

books collected on a desk for a project could always remember that they were stored together at a certain point on a certain day when a certain topic was being researched

a book could remember how long a reader spent on a certain page, as well as what other pages were read before and after (either in that book or in another source)

authors could be given raw or summarized data on reading and usage patterns for their work: how long people spent using certain sections, what other texts they also referred to at about the same time, and whether or not any of those texts were used in secondary work by readers

an article or Weblog post would know when someone else, somewhere in spimespace, read it, quoted it, or cited it (and the context in which it was read, quoted, or cited)

an in-progress draft could gather emerging new discussions about a relevant topic posted in other spaces (books, journals, online discussion spaces, Weblogs)

On one hand, this sort of discussion seems hopelessly utopian. On the other, stranger things have happened. I am not suggesting that we abandon printed books or handwritten notes or software. I am suggesting that we think about what we are doing in light of a different, still-emerging set of possibilities in order to identify positive and negative aspects and to discuss things we might want to try out.

Simple spime technologies involving RFIDs are already emerging (often driven by large organizations such as Wal-Mart, which use them for inventory control, and airports, which use them to track luggage). Furthermore, this emerging information sphere is, in many cases, driving economies in new directions, for example, sharing data using consistent formats such as XML for easier movement of information across organizational boundaries.

Other rudimentary forms of these capabilities are emerging, albeit in fragmentary and incomplete ways. The Technorati search (fig. 2.10) is one example. This view of the Technorati search is one I rarely see, because I have configured my account on the service to allow me to automatically gather new entries as they appear and display them in NetNewsWire, my RSS feed client, along with feeds from several hundred other Weblogs and Web sites. One part of this RSS feed includes saved Google queries on key topics I am researching so that as new Web sources emerge, I am notified of them in a single window on my screen. Similar spimelike capabilities are available in a wide range of sources—discussion boards routinely offer automated notification systems to alert users to responses to specific threads; and news sites such as Google News allow users to save keyword searches and be notified when news items are posted that include those keywords.[3]

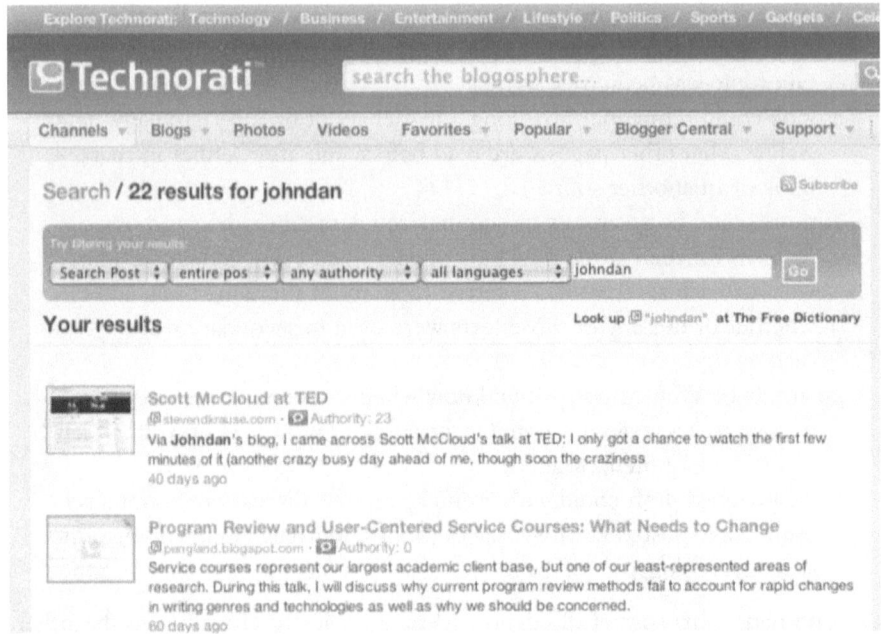

Fig. 2.10 Ego surfing via a Technorati search

The bridge between virtual and physical spaces is also suggested by products such as Delicious Library (see fig. 2.11) that provide a rudimentary way for organizing information about physical objects in a virtual space. Delicious Library relies on the Webcam called into service as a simple, handy UPC scanner. Users hold up a book, DVD, CD, or other physical item (anything with a UPC[4]), which is then scanned and used to download a product description and a digital image of the scanned item from Amazon's database. After gathering this information, Delicious Library users maintain databases of their texts and other materials, using the database assembled by users to track physical location (home or office), borrower, or other details. Delicious Library is not, obviously, completely spimelike, because it relies on a great deal of human intervention; the objects themselves are not tagged or otherwise self-aware. Scanning the UPC is a small, witty way to bridge a gap that spimes might eventually fill.

Development of such technologies does suggest the usefulness of even more-advanced technologies. Spimes at a micro (page) level would allow me to gather rough information about time spent per page by my large (okay, small) group of users, giving feedback about what is seen as useful, important, or offensive. I already have a limited version of this available using Web server logs to track users, but activities of my own texts in the print world—in other words, what readers are doing with the print texts I have written—are comparatively invisible to me.

Fig. 2.11 Scanning physical items into Delicious Library

Text spimes could also provide us with a platform for enacting database writing. Texts become assemblages of other texts: very leaky assemblages that can be "borrowed" from various sources and recombined with other text fragments without regard for clumsy, manual citation systems. The act of reading a text then includes the ability—something the Web and hypertext promised us but never delivered—to move from quotation to source to quotation, back and forth. Theodor Nelson worked out a system for all of this in the 1960s and 1970s called Project Xanadu, which unfortunately never really made it off the ground, leaving us with a relatively limited Web space that works great due to its simplicity—but is still hampered by its old-fashioned approach to writing and reading.

Most of us do not know who reads our work, unless readers initiate a conversation and tell us something they liked or disliked about a specific passage via face-to-face discussion, email exchange, or explicit citation that we happen upon in our own reading. How can we increase the conversation? In a fully spimed, textual world, other users' spime texts would automatically generate feedback if the users cited or referred to one's own texts (much as a Technorati RSS feed or a Google search does, but wider).

Readers of spime texts would also be involved in a more explicit ecosystem of textuality: Reading is no longer the invisible consumption of text but a semipublic performance, a form of distributed applause or hissing, less active than actually voicing a response but still an indication of something. Reading becomes transmissive and performative rather than simply privately receptive, at a functional level. And reading itself gains new status: a social action in addition to a cognitive one. Currently my most explicit and immediate evidence of how well my books are doing emerges in sales records, followed secondarily by crucial things like citation, quotation, and discussion. But if readers of spime texts are made more visible, they become more explicitly active and present. This does not discount the primary importance of discussion and use but provides a better bridge to those activities.

When spime functions are fully implemented, readers might be more encouraged to enter fully into conversations, writing their own texts to support their acts of reading.

## An Incomplete List of Concerns

If the spime text world described here seems obviously utopian, it also brings with it a fairly strong set of dystopian concerns. For example, although many users of the Web are not aware of it, most traffic on the Web is tracked in a multitude of ways, even outside of covert government surveillance. Services such as Google Analytics provide very simple-to-implement systems for Web site authors to track (in most cases) which specific computers access a Web page, including details such as the geographic location of the accessing computer and details about the computer (operating system, display resolution, and so forth), as well as the link that a user clicked on to arrive at a specific page. In many institutional settings, computers are often assigned semistatic addresses that make connecting this data to a specific user fairly simple.

For my own Weblogs, I use eXTReME and Google Analytics to track aggregate data such as how many users have viewed certain posts. The reports generated, though, also include items such as Google and Yahoo searches that lead users to my Weblog. Anyone who has skimmed though such search logs knows that there are a lot of people out there looking for some alarmingly bizarre things. I have avoided the temptation to track down identifying information about search engine queries ending up on my site, although doing so is certainly feasible.

Not surprisingly, tracking the movements of users in online spaces for surveillance reasons is increasingly common. A 2005 survey on electronic monitoring and surveillance by the American Management Association found that 76 percent of reporting workplaces monitored which Web sites employees visited; 26 percent of reporting companies had terminated workers for inappropriate use of the Internet.[5] And within the context of teaching, most courseware allows instructors to track such details as time spent by students looking at specific pages on the course Web site, the number of students who have accessed materials, and the length of time taken to complete some types of assignments. All of these activities may certainly have laudable pedagogical goals, but the ease with which they are adopted by many teachers suggests that not enough attention is being paid to issues of online privacy.

Fig. 2.12 Tracking who accesses Web sites via eXTReME (left) and Google Analytics (right)

Even more difficult are pedagogical issues that would normally encourage us to watch over our students as they work in careful and detailed ways. Consider, for example, how much more useful a portfolio would be if it allowed instructors to track a student's research in much more detail than currently afforded by the submission of notes, outlines, and drafts for commenting—to see what texts students read online, which they bookmark and quote from, how they arrange bits of information in various windows on-screen. Do they open multiple windows in ways that allow them to scan across diverse resources? Do they work on high-level issues such as rhetorical argument and arrangement before turning to surface-level revisions? When they are gathering facts from Web sites to use in an argument, do they click the "About Us" link to see who is behind the information?

Or to spin this in a slightly different direction, if you are reading my article, do I have the right to know that you have read it? That you have passed it on to someone else with a snarky comment? That you have quoted it in a text you are writing, agreeing or disagreeing with me? What ethical obligations does a reader have to an author? To the community? As Ben Vershbow observed, we are entering an era in which "the book is reading you."

Evaluating the ethical dimensions of contemporary technologies such as spimes, like evaluating computers or automobiles or vaccines, challenges us in productive ways. Although we long for the comfort of Google's famous corporate guideline, "Do no evil," the world is a complex place. Technologies are neither neutral carriers of intentions nor completely autonomous agents. Instead, technologies are articulations: ongoing, collaborative constructions being played out functionally, socially, and politically.

Are spimes good or bad? Yes. Both. Maybe neither. All of the above. Sometimes. The question is unanswerable without bracketing out so many aspects that the question itself becomes meaningless. They will be what we make of them, in particular instances and among particular sets of designers and users.

Spimes are, in addition to everything else, conceptual objects—moments of articulation that can, with some effort, help us call into question what we are doing and what we want to do.

### Notes

1. I am skipping an additional stage in Sterling's history: between artifact and product Sterling positions the machine. There are important technical and social reasons for including this stage in a wider account, but for texts, "machines" are, to some extent, not developed until the gizmo stage (but see the earlier discussion of Ramelli's bookwheel).

2. Only alluded to here is the immense amount of effort put into literacy education (individually, economically, and socially), which makes even this "straightforward" activity an immense barrier, even in "developed" countries.

3. At one point, my Bluetooth-enabled cell phone changed the Instant Messenger status on my laptop to "Available" when I was within ten feet of the computer or to "Away" when I was farther away. This eventually seemed like a little too much public location awareness.

4. Unfortunately, the UPCs on wine bottles are not unique to individual categories of products in the way that they are for things like books, so Delicious Library cannot yet be used to organize wine collections—ironic given that one of Sterling's main hypothetical examples in *Shaping Things* is a spime-enabled bottle of wine.

5. Law in this area is somewhat in flux: In 2008, the U.S. Ninth Circuit Court of Appeals found in favor of employee privacy when employees use third-party text-messaging services paid for by an employer (Morphy).

### Works Cited

American Management Association. "2005 Electronic Monitoring & Surveillance Survey." Accessed 23 June 2008, http://findarticles.com/p/articles/mi_m0EIN/is_2005_May_18/ai_n13726103/.

Basbanes, Nicholas A. *A Splendor of Letters: The Permanence of Books in an Impermanent World.* New York: HarperCollins, 2003.

Deleuze, Gilles, and Félix Guattari. *One Thousand Plateaus: Capitalism and Schizophrenia.* Translated by Brian Massumi. Minneapolis: University of Minnesota Press, 1987.

Foucault, Michel. *The Archaeology of Knowledge and the Discourse on Language.* Translated by A. M. Sheridan Smith. New York: Pantheon Books, 1972.

Godin, Seth. "Seth's Blog: Random Thoughts About the Kindle." Online posting. 19 June 2008. Accessed 24 June 2008, http://sethgodin.typepad.com/seths_blog/2008/06/random-thoughts.html#trackback.

Goodall, George. "Ramelli's Book Wheel." Accessed 23 May 2008, http://www.deregulo.com/facetation/2006/08/ramellis-book-wheel-book-wheel.html.

Graafstra, Amal. "Amal's RFID Implant Page." Posted 1 June 2006, accessed 27 May 2008, http://www.amal.net/rfid.html.

Greenfield, Adam. *Everyware: The Dawning Age of Ubiquitous Computing.* Berkeley, Calif.: New Riders, 2006.

Hitachi, Ltd. "Hitachi Develops a New RFID with Embedded Antenna μ-Chip." Posted 2 September 2003, accessed 23 May 2008, http://www.hitachi.com/New/cnews/030902.html.

Mohsin, Umair. "The Hunt." Posted 11 November 2007, accessed 20 January 2009, http://www.flickr.com/photos/umairmohsin/2067631039/.

Morphy, Erika. "Workplace Text-Messaging Ruling Wows Privacy Advocates." *TechNewsWorld*. Posted 20 June 2008, accessed 23 June 2008, http://www.technewsworld.com/story/Workplace-Text-Messaging-Ruling-Wows-Privacy-Advocates-63492.html?welcome=1214261458.

Morroia, Fabrizio. "Strahav Monastery Library." Posted 29 September 2005, accessed 23 May 2008, http://www.flickr.com/photos/biccc/47670267/.

Nelson, Theodor Holm. *Computer Lib/Dream Machines*. 2nd ed. Redmond, Wash.: Microsoft Press, 1987.

Rockwell, Geoffrey. "Text in the Machine." Posted 3 June 2007, accessed 3 June 2008, http://www.flickr.com/photos/geoffreyrockwell/528833325/in/photostream/.

Sterling, Bruce. "Dumbing Down Smart Objects." *Wired* (October 2004). Accessed 23 June 2008, http://www.wired.com/wired/archive/12.10/view.html?pg=4.

———. *Shaping Things*. Boston: MIT Press, 2005.

Tinderbox [Computer software]. Available at http://www.eastgate.com/Tinderbox/.

Vershbow, Ben. "The Book Is Reading You." Posted 19 January 2006, accessed 27 May 2008, http://www.futureofthebook.org/blog/archives/2006/01/the_book_is_reading_you.html.

# Serial Composition

Geoffrey Sirc

Here is the problem: since the first-year composition course began in the late nineteenth century, the primary instructional text, the expository essay, has remained the field's formal constant. To illustrate, we can use John F. Genung's influential 1892 text, *The Practical Elements of Rhetoric*, to limn the stock scene of academic composition. The program proceeds as follows: Selecting material is important, of course, but then comes the real work, "the business of building this material together into literary forms. . . . Out of the scattered elements at command is to be formed a structure of thought, which is to be no crude congeries jumbled together as it happens, but a unified, coherent, organic system. It is to such skilled combination alone that we can rightly apply the name style" (108). The process becomes an orchestration, in which various elements inflect both one another and the larger whole: "How words are related to one another grammatically; how they sound together; how they refer to what precedes or prepare for what follows; how their position is so to be determined as to give them force and distinction in themselves or make them a support to one another,—such questions as these arise at every step, questions to be answered only by constant and studious attention to the logical relations of the thought" (108). It is the theme, the "working-basis" of the work (248) that provides the determinant template for this part-to-whole orchestration; the theme "must be an idea so definite and clear-cut that the writer can resort to it for every step of his work. It is that nucleus-thought, expressed or implicit, which must be in his mind as a central point of reference, a constant determinator and suggester of the scope and limits of his subject" (248–49).

Once the theme is determined, the writer's task is set: namely, to "examine anew the various hints and shades of suggestion that lie involved in the theme, and systematize these into a plan of discourse, in which the accumulated material shall appear in properly subordinated, proportioned, and progressive sequence" (260). An outline or skeleton plan is constructed, to ensure that properly proportioned sequence, and then comes development of points and ideas, which Genung refers to as *amplification*. There are conventional means by which writers amplify their thoughts (enumerating the particulars of general statements, repetition, and illustration), as well as tools to use (quotation, allusion, suggestion). Editing and polishing follow, of course.

Such, in a nutshell, is the whole of college writing: generation of an essay theme, completion of a topic outline, and stock strategies and techniques for fleshing out the essay, making sure the parts all work together to inflect the whole, helping to produce that *properly subordinated, proportioned, and progressive sequence*. Over the years, about the only thing that has changed in college writing has been the amplification pattern offered students; so, for example, by the seventh edition of *Writing with a Purpose* (1980), McCrimmon offers the following under the rubric "Common Patterns of Development": illustration, comparison, classification, process, and definition (63–89). Roughly 120 years of college writing instruction have elapsed, with all the concomitant changes in culture and technology, and students in first-year composition work the same project in the same form with the same media.

Such has not been the case at all in other scenes with the idea of "composition" as a central focus: artists working in those fields (painting, sculpture, music, architecture, and so on) have understandably chafed under the constraints of established conventions and so created new formal strategies and genres; as those became established, further changes occurred. That dynamic of formal revolution was never mirrored in writing instruction. In this essay, then, I revisit a few of those other compositional scenes to see why, when other fields have changed in interesting ways, college composition has remained static. I choose a few scenes from the 1960s because that was an era when the discipline of writing instruction came closest to radical change, change that would have put college writing on a par with other fields doing intensive interrogation of the forms, means, and institutionalization of their compositions. In the 1960s, artists particularly wanted less ornately inflected styles, using simpler forms and materials, as an alternative to more complicated compositions made with relatively rarefied materials (whose uses one had to study formally to master). Today, with such pervasive technological mediation in our writing courses, we seem to be on another cusp; it is my hope that recuperating prior histories might offer productive reflection for contemporary practice.

### Scene 1: Primary Structures

In January 1963 there was a group show at the Green Gallery in New York City, one of a handful of key galleries then clustered together on the Upper East Side. The Green, which opened in 1960, quickly became known as one of the trendiest in town, always ahead of the curve for showing new art. In fact, the name "Green" was chosen for the gallery to imply newness. This particular 1963 show captured the spirit of newness through its very title, "New Work: Part 1." And among the artists exhibiting in the show were several that were soon to become dominant figures in the history of American art: Donald Judd, Robert Morris, and Dan Flavin. Those three, along with Carl Andre, Sol LeWitt, and Anne Truitt, were poised to launch a new movement on the art scene—minimalism— a movement that would hit the international community with amazing force in the next several years; so much so that the first major show of minimalist art, occurring three years later at the Jewish Museum, would be a black-tie affair, swarming with TV crews and paparazzi. That show, the title of which, "Primary Structures," fittingly captured the cool, neutral, reductivist aesthetic of the art, would earn incredible reviews: "A new aesthetic era is upon us," was how Hilton Kramer put it; "This year's Landmark Show," announced the *Times* (qtd. in Meyer 13). Although the excitement over the art is long past, the ideas of the minimalists persist today—not just as expressed in their art, but in their criticism, for they were among the first generation of artists to become known almost as much for their art texts as their art works.

Basically, the minimalists strove to further the abstract expressionist project: to establish a truly modernist art by ridding the artwork of all elements that were not exclusive to it. For minimalist sculptors, the best way to clear out the anthropomorphic illusionism of representational narrative and imagery was by reducing the work to a neutral object. And just as the modernist painters' focus on the materials of painting proved formally generative for them (that is, think of the effects Jackson Pollock achieved using enamel house paint, sticks, and glass basting syringes), so too the minimalist sculptors found that concentrating on matter rather than image changed the entire scene of their composition "from particular forms, to ways of ordering, to methods of production and, finally, to perceptual relevance" (Morris 67–68).

Robert Morris, in a 1966 manifesto of minimalist art entitled "Notes on Sculpture," reacting against perceptual ambiguities in work with "clearly divisible parts," noted that "simpler forms . . . create strong gestalt sensations. Their parts are bound together in such a way that they offer a maximum resistance to perceptual separation" (6). To achieve this strong gestalt, he urged a sculpture of "unitary forms," regular and irregular polyhedrons, which prevented complicated part-to-part or part-to-whole relationships from being established. But the viewer was warned against an easy dismissal of such simple shapes: "Simplicity of shape does not necessarily equate with simplicity of experience. Unitary forms do not reduce relationships. They order them. . . . They are bound more cohesively and indivisibly together" (8). SHAPE, then, was "the single most important sculptural value" (8). Minimalist work excluded details, such as colors, sensuous material, or interesting finishes, which Morris believed were "factors in a work that pull it toward intimacy by allowing specific elements to separate from the whole, thus setting up relationships within the work" (14). The idea, then, was not that highly inflected, complicated style of painting or sculpture in which major and minor themes orchestrate together to form a dense text. That kind of work had been done; rehashing it was *rétardaire*, resulting in work Morris dismissed as "candy-box art," "indulgently focused on surface, . . . coruscat[ing] with the minor brilliance of the '*objet d'art*'" (25).

Seriality became a key part of minimalist grammar because, as these artists realized, "Of all the conceivable or experienceable things, the symmetrical and geometric are most easily held in the mind as forms" (Morris 64). Seriality proved the perfect text grammar for a shape-oriented composition of primary structures. It was the most basic compositional method, simple parataxis or repetition—"one thing after another," as Judd put it (qtd. in Meyer 179)—a method in which there was no complicated, overarching compositional whole to keep inflecting (Meyer 171). Indeed, it was a method, not a style, according to the minimalist artist and critic Mel Bochner (*Solar System and Rest Rooms* 42), who notes, "Seriality is premised on the idea that the succession of terms (divisions) within a single work is based on a numerical or otherwise predetermined derivation (progression, permutation, rotation, reversal) from one or more of the preceding terms in that piece" ("Serial Art" 100). Bochner describes the serial grammar that serves as a modular system for Carl Andre's sculptures, pieces Bochner prefers to call "arrangements" rather than compositions, capturing the almost offhand, arbitrarily arrived-at nature of the work: "He uses convenient, commercially available objects like bricks, Styrofoam planks, ceramic magnets, cement blocks, wooden beams. Their common denominators are density, rigidity, opacity, uniformity of composition, and roughly geometric shape. . . . Only one kind of object is used in each [work]. . . . The arrangement of the designated units is made on an orthogonal grid by use of simple arithmetic means" ("Serial Art" 94). Witness, for example, Andre's piece in the "Primary Structures" show: *Lever*, a single row of 139 firebricks installed in a room with two entrances so that the viewer could stand at either entrance and have an unbroken, material vista, nicely capturing Andre's notion of sculpture as no longer form but rather "place" (qtd. in Bourdon 103).

So, the cube, brick, or metal plate—when these are serialized in a basic grid-type frame, generating a larger whole from initial bits or cuts, we have the morpheme and syntax of the minimalist text-logic. It is "the simplest ordering of part to whole," Morris claimed. "Rectangular groupings of any number imply potential extension; they do not seem to imply incompletion, no matter how few their number or whether they are distributed as discrete units in space or placed in physical contact with each other. . . . From one to many the whole is preserved so long as a grid-type ordering is used" (29). Minimalism was not an adversarial reaction to abstract expressionism, despite appearances—cool, cerebral, industrial forms rather than highly charged whipping and slashing of paint. Both of them were attempts to complete the modernist project of reducing the work to its essential elements. Morris revered Pollock as perhaps the most scrupulous investigator of forms and means. It was Pollock who defined the basic scene of composition: "tools, methods of making, nature of material" (Morris 44–45). But by reducing it, of course, he expanded it, as materials could now include sand, nails, keys, cigarettes, and aluminum fence paint. Drip painters such as Pollock and Morris Louis showed that compositional form was what *"resulted* from dealing with the properties of fluidity and the conditions of a more or less absorptive ground"; such forms were "not a priori to their means" (Morris 44, emphasis added). Composition itself became an investigation of means; otherwise, it was mere formalism. New materials needed to result in new forms.

There is an elemental poetry in minimalism, and so one thinks, perhaps, of Wallace Stevens's jar on the Tennessee hill, equally minimalist, "gray and bare," ordering relationships in the modest form of the anecdote. Or John Ashbery, whose method was to find the poetic possibilities in ordinary language, and who conceived of each of his poems as an empty oblong box—a primary structure, a minimalist form—and filled them with cuts from a variety of material sources: "people talking, journalese, pop culture, cracker-barrel philosophy, high-flown poetic diction" (MacFarquhar 92).

### Scene 2: Kitzhaber/Andre

In 1963 the field of composition had its own "landmark show," the publication of Albert Kitzhaber's book *Themes, Theories, and Therapies: The Teaching of Writing in College*, the first major critical examination of college writing instruction in the modern era. Kitzhaber's book was actually the report of the Dartmouth Study of Student Writing, a study funded by the Carnegie Corporation undertaken to examine why professors at Dartmouth were less than enchanted with the writing abilities of their students, especially given the two-quarter sequence of composition nearly every freshman had to take. The central question the study was designed to determine was "Can English composition at Dartmouth be taught more effectively?" (ix).

Clearly, the study was a daunting task. As Kitzhaber notes early on, there are all sorts of curricular ways to foster growth in writing in a ten- or fifteen-week period, "as long as the students are reasonably normal and are doing some writing under supervision" (4). Tracing exactly how or whether a given intervention strategy helped a student is almost impossible. It is also a stretch to think that any one- or two-course sequence can foster substantive growth in writing, can, in Kitzhaber's words, "develop a well-stocked mind, a disciplined intelligence, and a discriminating taste in language and fluency in its use. None of these can be acquired without hard work over a period of years" (7). As part of the project,

Kitzhaber and his assistants studied 495 essays from a cross section of Dartmouth's first-year literature-based composition classes; they "classified and recorded all errors, infelicities, weaknesses, and other negative criticisms that the teachers had noted on the papers" (42), from errors in focus down to those in diction and spelling. One of the first things Kitzhaber noted was that even when the course parameters narrowed the focus to Milton and Shakespeare, instructors still permitted students to write more informal, personal essays on topics like beatniks and school spirit. Astonishing, too, was the wide variety in marking and grading. Most instructors overmarked: as many as seventy-five errors might be highlighted in a three-page paper. As Kitzhaber describes such instructors, "A misused semicolon or an off-center idiom afflicts them like an uncontrollable itch, and they are not comfortable again until they have scarified the error with a red pencil" (58–59). Some undermarked, "placing three or four marks in the margin, a gnomic comment at the end, and a C⁻ at the top" (59). Some corrected papers by simply using "rule numbers and cryptic abbreviations" (65) keyed to the handbook code. Kitzhaber felt particularly bad about the student who got a paper back covered with such coded symbols and only a single written comment: "You misspell 'Shakespeare'—for shame!" (65).

In his study's final recommendations, Kitzhaber addresses many of our as-yet unresolved issues—staffing, teaching load, instructional focus, assignment design, course content, and grading policies. His most interesting suggestion, however, is a modest (too modest, perhaps) plea for variety in the genres students should work through in order to become better writers. He suggests a host of small assignments, such as expanded definitions, close readings of a passage, parodies, and analyses of style and structure, to augment the series of analytic, expository essays that will be the student's main focus in the course. I say "too modest" because Kitzhaber earlier complains about how "nothing is being done in the colleges to reform the freshman course. There is no widespread impulse to think through afresh the premises and purposes of this course (or perhaps one should say to think them through for the first time)" (98).

Minimalism was probably too new for Kitzhaber to be aware of at the time—a shame, because the writings of minimalist sculptor Carl Andre would have provided interesting options for the textual variety Kitzhaber advocated in first-year composition. Andre was like many of our students when it comes to writing—uncertain, resistant: "I have never been a writer of prose," he claimed. "I have never felt comfortable writing prose; it is something that is very difficult for me. . . . My own mind moves by no means of prose" (3, 125). Hence, many of the texts Andre wrote, as his editor, James Meyer, points out, were less than a hundred words long. Unlike our students, though, he had no strictures against sustained work in forms where brevity was an option. Meyer (4–11) provides a taxonomy of Andre's preferred genres: the *statement*, generally fifty words or less, crafted for an exhibition; the *dialogue*, the record of a written or oral interview, in which questions prompt reflections, which prompt more questions, a perfect form for a resistant writer, in the way "writing begets more writing" (5); the *epistle*, letters of varying lengths to varied recipients—another attractive genre for a resistant writer, as such correspondence "may be informally composed or carefully wrought, a lengthy missive or a postcard" (6); *epigrams* and *maxims*, terse, witty, insightful statements containing "few asides, parenthetical remarks [or] dependent clauses" (7), often in the form of chiasmus or syllogism; and his *planes*, experiments in planar poetry, where words are fugued together in a grid pattern.

Such a concept, refiguring the curriculum around shorter assignments, allowing student resistance to be the engine of student writing, is doubtless a tough sell. A recent report on undergraduate learning at the University of Washington, for example, found that "first-year students generally find shorter papers easier to write than longer ones, and they often do not spend much time or effort writing papers that are fewer than four pages long" (Beyer, Gillmore, Baranowski, and Panganiban). One can only wonder at the assignments UW students were given in those short papers, whether they truly attempted to develop the pithy, the poetic, or whether the assignments were simply less challenging opportunities to practice the prosaic.

The writing class, then, as a space where students primarily learn writing through a series of conventionally organized essays had not, in 1963, changed much at all since the field's nineteenth-century origins. In Dartmouth's first semester course, it was seven themes of eight hundred words; the second course entailed three more of those essays plus a research paper of about two thousand words. In writing initiative after writing initiative, we have never really questioned the strategy of teaching the essay by having students write essay after essay, despite persistent disenchantment with the essays that get written. Andre believed that "any task can be accomplished if you divide it into units small enough" (275). We too often offer students a curriculum in which "quality" is the key criterion, a questionable goal, perhaps, for learning the craft of writing in the first year. "Whenever you see or hear the word 'quality' in art, understand the word 'commodity' is meant," reads one of Andre's maxims (30).

Andre's textual genres are focused on the material stuff of language, on words. No surprise—when he was growing up, the dictionary was his family's bible; his mother was an amateur poet and his immigrant father loved to come home with new words he had learned at work, springing them on his family, after which they would look up their etymology. Words became his textual emphasis, much like the individual material unit, whether brick or metal plate, had primacy in his sculpture. Meyer sums up Andre's method: "He developed a *nonsyntactical syntax* that stresses the part (the 'cut') rather than the whole. Where the old syntax is predicated on an established, a priori grammar, the new syntax is based on the *unit*'s grammatical potential. The work's form is continuous with its internal elements—their shape, their density, their size. The relation is no longer that of part to whole, but of whole to part" (12).

Andre's writing, then, is a series of "cuts," strung together by a minimalist grammar of seriality. Kitzhaber's flaw, a persistent one throughout the history of writing instruction, is maintaining an unquestioning insistence on aestheticizing the form of first-year composition as the thesis-driven college essay, where all elements inflect together in that critical part-to-part/part-to-whole emphasis so as to cohere into an autonomous work. There was no revision of the object of practice in Kitzhaber's critique of the scene of first-year writing, just the way it was taught. In my analogy, it is as if the formal structure of cubism has never been ruptured in writing instruction. In the visual arts, modernism meant, in part, questioning that highly determined cubist program, with its division of the picture-plane into a carefully planned, organically articulated structure. Pollock was arguably the most influential American artist in history, for his experimentation with form, process, and materials, casting aside the academic tradition's "painterly-artistic elements," to use Kasimir Malevich's words for the traditional conventions then haunting the scene (qtd. in Morris 51). The sculptor Richard Serra summed up the influence of Pollock's bold disregard in noting that he was "not playing the same game as Vermeer": "We evaluate artists by how much they are able to rid themselves of convention, to change history. Well, I don't know of anyone since Pollock who has altered the form or language of painting as much as he did" (qtd. in Kimmelman). Andre wanted more *kunsthalles* in America, museums with no permanent collections, so as much work as possible could be seen and appreciated, without the commodification of art and reification of quality that comes from its institutionalization in a collection. Dartmouth students, when the compositional world was changing, and new materials made new forms possible, were stubbornly judged by the prescribed forms of a traditional aesthetic. Forms never interested Andre as much as materials. He loved matter, the properties of stone or metal or wood or hay. He combined them but never joined them permanently with welds or rivets, so as to preserve their quality as pure cut as much as possible, rather than transmute them through an arbitrary, predetermined combinatory logic. A serial grammar was ideal for a focus on the actual stuff of the compositional unit in the way, as Frank Stella noted, the use of repetition "drew attention to 'the thing itself'" (Meyer 169–170). The art critic David Bourdon traces the origin of Andre's method.

> From 1960 to 1964 Andre worked as a freight conductor and brakeman for the Pennsylvania Railroad in Newark. Though he had already begun to work with preexisting, standardized materials, four years of coupling and uncoupling freight cars confirmed him in his use of regimented, interchangeable units. Because any part could replace any

other part, the materials did not lend themselves to relational structures. In refusing to determine the mutual relations of forms, he suppressed his desire to compose. (104)

Our current insistence on throwback notions of expository analysis is tired not just stylistically, but rhetorically. We cling to the received notion that the ultimate goal of a given essay is to convince or persuade a reader. Longinus, the classical theorist of the sublime, would die laughing. What he was after should be the goal for every writing class: excellence in discourse, figuring out how "the greatest poets and prose writers have acquired their pre-eminence and won for themselves an eternity of fame," and it was certain to him that they never won it by bothering to persuade anyone of anything.

> For the effect of elevated language is not to persuade the hearers, but to amaze them; and at all times, and in every way, what transports us with wonder is more telling than what merely persuades or gratifies us. The extent to which we can be persuaded is usually under our own control, but these sublime passages exert an irresistible force and mastery, and get the upper hand with every hearer. Inventive skill and the proper order and disposition of material are not manifested in a good touch here and there, but reveal themselves by slow degrees as they run through the whole texture of the composition; on the other hand, a well-timed stroke of sublimity scatters everything before it like a thunderbolt, and in a flash reveals the power of the speaker. (114)

Even more than catharsis, then, the goal is ecstasy. "All art everywhere all the time," Andre advocated in one of his statements (30).

## Scene 3: Cassette / Mix Tape

In 1963 the world outside the college classroom kept turning. That was the year, for example, that the Royal Philips Company of the Netherlands unveiled a new product, the compact audiocassette. It took a few years to perfect the mechanism, but by 1966 the new audio format had been standardized. Not until 1979, of course, when Sony introduced its Walkman, did the cassette became a hugely successful, transformative part of the cultural landscape. As a result, James Paul, critic for London's *Guardian*, claimed, "Our relationship with music has never quite been the same since. . . . It allowed us to listen to music differently, privately. And it brought out the librarian in us: listing, labeling, indexing. With LPs you collected music; with cassettes you possessed it."

Once you owned it, you could refigure it any way you wanted; hence, the birth of the mix tape, a serial form drenched in a collector's obsession, termed by Luc Sante "a paradigmatic form of popular expression" (22). To demonstrate, here is a snapshot from mix tape history, courtesy of Thurston Moore, then-dishwasher, soon-to-be guitarist for the postpunk band Sonic Youth:

Around 1980–81, there was a spontaneous scene of young bands issuing singles of super-fast hard-core punk, most of which subscribed to a certain formula of thrash. Bands like Minor Threat, Negative Approach, Necros, Battalion of Saints, Adolescents, Sin 34. . . . They were great live and they made really great records. Very on-the-cheap and each tune was hardly a minute long. I was fanatical and bought them all as soon as they came out. I would stop each day at the Rat Cage on Avenue A and buy any new hardcore 7" they'd have on the wall. . . . But I also felt I needed to hear these records in a more time-fluid way, and it hit me that I could make a killer mix tape of all the best songs from these records—and since they were all so short and they all had the same kind of sound and energy, the mix would be a monolithic hardcore rush. . . . I made what I thought was the most killer hardcore tape ever. I wrote "H" on one side, and "C" on the other. That night . . . I put the cassette on our stereo cassette player, dragged one of the little speakers over to the bed, and listened to the tape at ultra-low thrash volume. I was in a state of humming bliss. The music had every cell and fiber in my body on heavy sizzle mode. It was sweet. (10)

Given the mix tape's status as highly prized alternative text, mix tape artists search for and theorize "the perfect mix tape." Their strategies in that search form a post-album-format "elements of style." Jack Tripper, on the Tiny Mix Tapes site, justifies the mix tape as compositional genre, differentiating it from the randomness of shuffle technology: "It takes enormous amounts of outlining and planning before executing a perfect mix tape. Sure, you can throw a bunch of random songs together, but don't come crying to Jack Tripper when that special someone dumps you or your new best friend ditches you—because they will. I promise you, if you follow these little guidelines, you'll have that special someone or best friend for at least a month longer."

First, you need a theme. A host of standard mix tape themes have evolved: the romantic mix, the breakup mix, the friendship or platonic mix, the intro-to-genre-X mix, the road-trip/airplane mix, and the party mix. To those stock themes, some have added the workout mix, the ambient mix, the sleep mix, the hangover mix, the alphabetic mix, and the mix of all cover songs. But, of course, in a genre where the coolly sublime is highly prized, we find even more outlandish themes showcasing a mix taper's style and wit: "cleaning up after the party" mix, "lost my damn job" mix, "scare your neighbors" mix, "being laid up sick in bed for two weeks" mix, "i really wish i was a pirate" mix, "for my pets while i'm not at home" mix, "can't understand a word they are saying" mix, "no song more than 30 seconds" mix (*Art of the Mix*); even the "songs whose 'titles would make awesome T-shirt slogans'" mix (Wilson).

After your theme comes the tracklist selection. About song choice, Tripper cannot overstate: "Don't throw on any shitty songs. I don't think they will appreciate listening to shit." Music critic Sara Bir describes the desired skill: "If you get the right flow going, it's possible to move from Donovan to the *My Fair Lady* soundtrack to Wilco without losing continuity." This tracklisting, which Tripper considers the "single most crucial aspect of mix taping" should be painstakingly experimented with and reviewed because, after all, Tripper adds, "Radiohead almost broke up over the tracklisting for *Kid A*."

The first song in the playlist is the toughest; it cannot be "obvious, cheesy, or predictable. And it can't be too obscure" (Tripper). A good first-song choice can act as the genesis for the rest of the tracklist. Also important are transitions between songs, especially if your theme calls for a variety of genres: "You can't just go straight from pop rock to detached, experimental post-rock. You need a link. Come up with songs that may fit in between, and if you can't find any, then one of those songs has to take a hike" (Tripper). *Globe & Mail* critic Carl Wilson describes how he "learned to finesse transitions: same key, new speed; same tempo, new key; startling counterpoint; found-sound bridge; chill down; epic climax; quick comic coda. . . . I would build narrative arcs, Socratic dialogues between, say, Billie Holiday and the Pixies." Nick Hornby offers one of the most oft-quoted mix tape directives in his novel *High Fidelity*:

> A good compilation tape, like breaking up, is hard to do. You've got to kick off with a corker, to hold the attention . . . , and then you've got to up it a notch, or cool it a notch, and you can't have white music and black music together, unless the white music sounds like black music, and you can't have two tracks by the same artist side by side, unless you've done the whole thing in pairs, and . . . oh, there are loads of rules. (89)

What the mix tape offers composition is proof of how a minimalist citational logic can achieve maximum ideational effect. The tracklist, especially when combined, as often happens, with the gloss of a mix taper's commentary for each song, becomes a kind of "primary structure" or "unitary form" grammar: the cut, and the comment, in serial order. Denial of the significance of such a popular form—both as learning-tool and as an end-text in itself—seems cranky and wrong-headed. All the complex painterly-artistic elements and part-to-whole inflections of the essay are wonderful, but hyping that genre so exclusively to students—as Kitzhaber & Co. did in 1963, cassette tape lurking in the wings, "New Work: Part 1" appearing in the gallery—is asking students to believe in the ordered logic of the album, to re-constitute the pre-recorded CD as serviceable form. And that, in the era of digital downloads, is impossible.

The mix tape ruptures the text of the accreted past. To possess the music is to be able to re-stage the scene of its fate. Longinus saw it coming: "the true sublime uplifts our souls; we are filled with a proud exaltation and a sense of vaunting joy, just as though we had ourselves produced what we had heard"

(120). A fitting comment, since Longinus's textual method is to re-stage the scene of his own favorite cuts, offering a mix tape of quotations as a treatise on rhetoric. Anne Carson describes Longinus's achievement in *On the Sublime:*

> "You will come away from reading its (unfinished) forty chapters with no clear idea what the Sublime actually is. But you will have been thrilled by its documentation. Longinus skates from Homer to Demosthenes to Moses to Sappho on blades of pure bravado. What is a quote? A quote (cognate with *quota*) is a cut, a selection, a slice of someone else's orange. You suck the slice, toss the rind, skate away. (45)

The critic Matias Viegener, likening mix tapes to the cento and avant-garde forms like the cut-up, Xerox art, and sampling, appreciates this documentarian banditry, the way mix tapes make "the existing world tell tales it does not intend to tell. You get the world to send you a message it never meant to send" (qtd. in Moore 35).

I am uneasy, certainly, with the way a simple minimalist text-logic like the mix tape can, with all its finessed transitions and Socratic dialogues, become just as highly determined as the most belletristic essay. At its most accessible, though, the mix tape is the product of a simple, linear combinatory logic, its thematization acting as the sole inflective principle, much like a chord or scale acts as the sole determinant for a minimalist composer. The theme enables that building of structure through the repetition of discrete cells that served as the necessary criteria for composers of minimalist music, where the end result was that genre's slow, harmonic build (Moore's "state of humming bliss"); it is text as audible structure. The most Longinian thematization-principle, possibly, becomes DJ Yoda's "I record anything I think is cool."

And then, of course, there is this: In 1992 Maxell sold 350 million blank audiocassettes, according to its vice president, Peter Brinkman; ten years later that figure was 140 million. Projections are that by the end of this decade, the audiocassette format will be obsolete (Stuever). This evokes in some the elegiac, the fetishistic. Sara Bir, who has grudgingly made the switch to mix CDs, writes longingly of the cassette medium, "as beautiful in its hiss as medieval manuscripts are in their decay." "It seemed enchanted," Carl Wilson writes. "CDs and iPods can't match the Proustian pungency of the cassette—Dolby hiss, Crayola scent, brittle weight in hand, paper, marker, glue." Mourned, too, is the defining limit of the audiocassette's side A / side B format, which determined a different kind of textual logic than the simple storage-space limit of a CD (not to mention the seeming limitlessness of an iTunes playlist). The *Washington Post*'s Hank Stuever calls the cassette's two sides a "crucial dialectic." Bir writes of them as "allowing for a first act, an intermission and a second act."

But whereas the cassette may be dying as a medium, the mix tape is certainly robust as a genre, thanks to MP3 audiofiles, the ability to download them, the click-and-drag ease of fashioning them into playlists, and—with burning technology—the simplicity of making a mix CD. Purists, of course, bemoan the craft, the personalism of the hand, and hence, the aura, the presence, missing in the digital mix tape. University of Wisconsin journalist Chris Vinyard, typifying this retro snobbishness, sneers at the mix CD: "It can be argued that the ease with which these mix CDs are made has taken away from some of the genuine qualities behind the mixtape and music sharing. Whereas in the past, one might have spent hours next to a tape recorder while recording every song in real time, 80 minute CDs are burnt in a few minutes flat. Now that mixes are so easy to create, there are more of them being made than ever."

"Thoughtless tune dumping" is how Carl Wilson terms it. This is the academy's anxious suspicion about any composition, when a supposedly painstaking process to master can be dashed off so easily. Joel Keller, writing in *Salon*, for example, after detailing his own hours-long process, including getting recording levels right and finding just those perfect-length songs that would take him to the exact end of a tape, mourns this loss of mystic presence (now termed "connection"): "The process of making a mix tape gave people a connection with music that the electronic version can't replace. Because it is so easy to drag and click a mix into existence, the sense of satisfaction with making what many feel is a work of art gets diminished." And, when the craft and aura are emptied from the scene, quality and discernment go as well; and so he sniffs, "Fewer people who are connected to the music they listen to translates into a less critical and picky audience for the crapola that the record companies and radio stations promote. The quality of music overall goes downhill."

It is like hearing mandarins scoff at the first photographs, denying them art status because the craft and presence of the painterly were gone; the camera, too, democratized composition, with the same point-and-click ease. But, far from being easier and more thoughtless, the digital mix tape is vastly more complex because now a tracklist is created not just from one's own record collection, but countless other available MP3 files offering an incredible range of material. The Longinian cut, then, becomes an even more crucial compositional value. And Andrew Leonard, an editor at *Salon*, claiming the technology has "helped usher in a renaissance of mix-tape brilliance" (indeed, he feels the making of mix CDs is what computers and the Internet were made for), sees the technological ease in the making of the actual mix CD as allowing for increased focus on material selection: "more time to pay attention to what really counts: the music. More time to be a perfectionist with regard to the essence of a compilation—the act of song selection." "The hardest thing in art," Andre writes, "even before you find your limits, is to find that which pleases yourself" (32).

The digital age has also evolved an interesting variant of the mix tape, the MP3 blog. Combining both musical selections and commentary, they have been described as "a slow-motion mix, a mash note to readers" (Wilson). They literalize, in a sense, the grammar of the typical blog, in which the postings so often consist of cuts from another source with the blogger's reflections. The blog, then, with its serial grammar of the cut and the commentary, acts as a textual interzone between the popular form of the mix tape (as catalog of pure cuts) and the academic essay, with its rejection of seriality in favor of a highly inflected arrangement of analytic exposition. The MP3 blog combines the time-fluid contingency of the mix tape with the canonical autonomy of the essay. The minimalist work, in its large, simple scale, was more public than intimate. So, too, the rectangularly grouped blocks of blog text are meant for a highly public readership (witness how often blog text is culturally recirculated). What typified the new sculpture of the 1960s for Morris was that its "order is not based on previous orders, but is an order so basic to culture that its obviousness makes it nearly invisible" (27). One does not worry in a blog about such "painterly-artistic elements" as a strong opening, a clear thesis, sufficient development, or a clever conclusion. More often than not, well-chosen cuts and a few lines of interesting commentary will suffice. Blogs, then, support the minimalists' claim that seriality, whether regular or irregular, can bring to material a de facto cohesiveness, because the viewer brings a way of reading that looks for "significant clues out of which wholeness is sensed rather than perceived as an image" (Morris 61). Sometimes a bunch of hard-core 7" LPs are all you need for a really rich text.

College Writing might not care that both the production and reception sites for text are changing so rapidly, but the rest of the world does. Ironically, writing teachers maintain an allegiance to a nineteenth-century essayist program when, today, the very nature of reading is drastically morphing. Cultural critic Nicholas Carr describes the change: "Immersing myself in a book or lengthy article used to be easy. My mind would get caught up in the narrative or the turns of the argument, and I'd spend hours strolling through long stretches of prose. That's rarely the case anymore. Now my concentration often starts to drift after two or three pages. I get fidgety, lose the thread, begin looking for something else to do. I feel as if I'm always dragging my wayward brain back to the text. The deep reading that used to come naturally has become a struggle" (57).

Carr knows the cause: It is the way so much of his textual life has switched to the computer screen, particularly Internet-based reading, the "universal medium" for textuality, as he terms it, "the conduit for most of the information that flows through my eyes and ears and into my mind.... What the Net seems to be doing is chipping away my capacity for concentration and contemplation. My mind now expects to take in information the way the Net distributes it: in a swiftly moving stream of particles. Once I was a scuba diver in the sea of words. Now I zip along the surface like a guy on a Jet Ski" (57). Textual technology has changed text processing; Carr speaks of people who could once handle lengthy texts like *War and Peace* easily, but now are lucky to get through three or four paragraphs of text online. A serial composition of short, staccato bursts seems essential as a compositional strategy for our age. This is exactly the kind of writing found on MP3 blogs. Here is Matthew Perpetua, from the audioblog site Fluxblog. He has posted the MP3 file of 'Accidental,' a track by Inara George and Van Dyke Parks, to which he appends the following brief commentary: "Inara George & Van Dyke Parks 'Accidental'—Van Dyke Parks' arrangement is in constant motion—swirling, twirling, dancing off in tangents. Nevertheless, the piece feels strangely static, as if Inara George's whimsical reverie was confined to a very small space, like a large scale musical theater production in a studio apartment. George comes across like a neurotic young woman wishing herself into the role of the romantic ingénue, and largely succeeding despite an inability to shake off her anxiety, or totally dial down her bitterness." This is exposition for that distracted reading Carr describes. Perpetua has mastered the style of offering just enough relevant detail and theme to convey an impression of the song. The reader listens to the track, and either agrees or disagrees with Perpetua, but in any event can appreciate the substance in his on-the-fly critique. Then, the reader hops back up on the Jet Ski.

This style of MP3 commentary has at least one precedent: the brief gallery reviews published in the 1960s by such minimalist artists as Donald Judd and Mel Bochner, who lent their textual talent and critical sensibility to art publications at the time in order to supplement their income. Here is one of Bochner's reviews, this of a December 1965 show of James Hans's work—"James Hans: Hans throws the works at every picture. They burst with eclecticism, as if in homage to a mythical god of art magazines. Photographs, photostats, collages, impasto, drip, 'fool-the-eye,' copied bits of Van Gogh, etc., only serve to deaden the viewer despite the sense of bravura that Hans certainly displays" (*Solar System* 1).

Pithiness, *le mot juste*, telling metaphor—these are the new elements of Internet-based style we have to teach. And even though Carr seems to pose the change in reading habits as dichotomy (either Tolstoy or Google), the effect of serial style—short, well-chosen bricks of meaning combining to form a rich whole—means we do not need to value brevity at the expense of that complexity of meaning traditionally thought to be available only through the studiously inflected part-to-whole thematized exposition of essayist prose. *New York Times* media critic Virginia Heffernan writes about the phenomenon of everyday viewers posting comments in response to videos they watch on YouTube. In particular, she focuses on the video "The Truth about Islam from an Ex-Muslim Lady," which at the time (November 2007) had prompted the most comments

(200,000) of any video on the site. Heffernan demonstrates the power of serial logic when she describes the cumulative effect of one brief comment after another, the way what might seem like a "ceaseless shouting match" actually coheres into a thoughtful text: "Part atavistic race riot, part religious disputation and part earnest effort at enlightenment, the expansive commentary is fast becoming a full-blown novel of world religion, one that dramatizes the fascinating and often shocking preoccupations of today's desk-chair ideologues" (23). Juxtaposition creates its own dense meaning.

Longinus's simple *ars rhetorica*, the stringing together of cut-and-comment, becomes the simplified text-logic of writing in an expanded field. Carl Andre has claimed, "The forms of my work have never particularly interested me. What has always been my search really is for material, a particle of material. It's finding a material or unit of material like a brick of the right size and the right shade and density and so forth—from finding this particle, I would combine it with others to make a work. I never in my mature work start with a form, a completed form. . . . The origin of the works has been finding things in the world to combine" (99).

## Conclusion

I worry that we are replaying the panic scene of Comp '63: a refusal to reduce, to empty out; no trace of minimalism's "cool" aesthetic of refusal; rather persistent faith in traditional forms and materials; an insularity, still, from both the theories and practices of high art as well as the forms and desires of the popular. I am advocating the possibilities of reduced geometries in our conceptions of composition. Writing not as a highly detailed system or grammar but as simply a practical field. A nonsymbolic approach, focusing on the materials involved and their basic combination. The contingent, rather than the canonical; reconstituting banal objects and juxtaposing them in interesting ways. I am betting students can learn more about writing from iconological projects named "Driving in Cars While Smoking," "Seduction Theory," or "Oral Surgery Disasters"—to cite the titles of some of the cherished mix tapes in Thurston Moore's collection—than from analytic exercises such as "The Purpose of a Hobby," "The Unskilled Worker and Life," or "A Comparison of the Resurrection of Christ and Pagan Resurrection Myths," a few Dartmouth theme topics. A textual goal far more fruitful in the first-year composition class than *quality*, which "is judged by reference to the standards not only of the old masters but of the great moderns, . . . an encomium bestowed upon aesthetic refinement," is *interest*, "an avant-garde term, often measured in terms of epistemological disruption" and which can "license critical inquiry and aesthetic play" (Foster 46).

Brian O'Doherty's insight about the minimalist sculptors—that they were not interested in making art, "just making" (253)—provides a way to refigure the composition classroom, shifting its focus away from a highly determined, overly prescriptive formalism and onto the simpler idea of making form. The need for first-year composition to emphasize the making of different forms is especially salient now in this era of expanded means and materials. Morris ends the final part of his "Notes on Sculpture," post-Pollock, post-Duchamp, with a reclamation of process at the expense of the iconic, finished form, which he terms "the craft of tedious object production" (68): his definition of art in that piece is "mutable stuff which need not arrive at the point of being finalized with respect to either time or space" (68). Morris, then, preserves and extends Moore's need for time-fluidity in composition (the MP3 playlist is always being tweaked). The art of the present, as Morris wrote in 1966 (anticipating the digital text as an in-process series of evolving iterations), is characterized by impermanence; a "conclusion" can be forced on it only by "'freezing' it into a static form." As indeterminate, contemporary work "can have any number of 'records'—the work itself does not come to rest with any of them" (69). The operative gerund for writing becomes *forming*. The criteria for a minimalist aesthetic of text-as-sculpture might include: symmetry, nonhierarchic distribution of parts, general wholeness, openness, extendability, accessibility, immediacy. Text as simple ordering of whole to part; cuts loosely assembled, rather than glued or riveted, so material can be prised out and linked again; with the parts as interesting as possible; a low-boredom writing; the ultimate goal—producing that Moorean (or is it Longinian?) monolithic hard-core rush. The blog, then, as essay manqué. In fact, the operative principle of the blog is captured in the very title Morris used to sum up his aesthetic of change, paradox, and rupture: *Continuous Project, Altered Daily*. "To whom is the artist responsible?" Andre was asked in a 1976 interview; his answer, "To the values of a craft—a process of making and selecting—and to the task of making that craft intersect with contemporary life as it is felt and seen" (40).

## Works Cited

Andre, Carl. *Cuts: Texts 1959–2004*. Edited by James Meyer. Cambridge, Mass.: MIT Press, 2005.

*Art of the Mix*. Accessed 23 July 2008, http://www.artofthemix.org/writings/history.asp.

Battcock, Gregory, ed. *Minimal Art: A Critical Anthology*. New York: Dutton, 1968.

Beyer, Catharine H., Gerald Gillmore, Matthew Baranowski, and Naomi Panganiban. *Writing at the UW: The First Year*. Report. University of Washington, Office of Educational Assessment and Office of Undergraduate Education, February 2003, http://www.washington.edu/oea/pdfs/reports/OEAReport0303.

Bir, Sara. "Mix Emotions." *Metroactive*, 22 June 2005. Accessed 23 July 2008, http://www.metroactive.com/papers/sonoma/06.22.05/mixtapes-0525.html.

Bochner, Mel. "Serial Art, Systems, Solipsism." In Battcock, 92–102.

———. *Solar System and Rest Rooms: Writings and Interviews, 1965–2007*. Cambridge, Mass.: MIT Press, 2008.

Bourdon, David. "The Razed Sites of Carl Andre." In Battcock, 103–8.

Carr, Nicholas. "Is Google Making Us Stupid?" *Atlantic*, July/August 2008, 56–63.
Carson, Anne. "Foam (Essay with Rhapsody): On the Sublime in Longinus and Antonioni." In *Decreation: Poetry, Essays, Opera.*, 43–57. New York: Knopf, 2005.
Foster, Hal. *The Return of the Real*. Cambridge, Mass.: MIT Press, 1996.
Genung, John F. *The Practical Elements of Rhetoric*. Boston: Ginn & Co., 1892.
Heffernan, Virginia. "God and Man on YouTube." *New York Times Magazine*, 4 November 2007, 22–23.
Hornby, Nick. *High Fidelity*. New York: Riverhead Books, 1995.
Keller, Joel. "PC's Killed the Mix-Tape Star." *Salon*, 22 Jan. 2004. Accessed 22 January 2009, http://dir.salon.com/story/tech/feature/2004/01/22/mix_tape_one/index.html.
Kimmelman, Michael. "A Drip by Any Other Name." *New York Times*, 12 February 2006, sec. 4.
Kitzhaber, Albert R. *Themes, Theories, and Therapy: The Teaching of Writing in College*. New York: McGraw-Hill, 1963.
Leonard, Andrew. "Praise Be to the CD Burner." *Salon*, 22 January 2004. Accessed 22 January 2009, http://archive.salon.com/tech/feature/2004/01/22/mix_tape_two/index.html.
Longinus. "On the Sublime." *Classical Literary Criticism*. Translated by Penelope Murray and T. S. Dorsch. London: Penguin, 2000, 113–66.
MacFarquhar, Larissa. "Present Waking Life." *New Yorker*, 7 November 2005, 86–97.
McCrimmon, James. *Writing with a Purpose*. 7th ed. Boston: Houghton Mifflin, 1980.
Meyer, James. *Minimalism: Art and Polemics in the Sixties*. New Haven, Conn.: Yale University Press, 2001.
Moore, Thurston. *Mix Tape: The Art of Cassette Culture*. New York: Universe Publishing, 2005.
Morris, Robert. *Continuous Project, Altered Daily*. Cambridge, Mass.: MIT Press, 1993.
O'Doherty, Brian. "Minus Plato." In Battcock, 251–55.
Paul, James. "Last Night a Mix Tape Saved My Life." *Guardian*, 26 September 2003. Accessed 23 July 2008, http://www.guardian.co.uk/music/2003/sep/26/2.
Perpetua, Matthew. "Where's the Knife? Where's the Fire?" *Fluxblog*, 23 July 2008. Accessed 5 January 2009, http://www.fluxblog.org/2008/07/wheres-the-knife-wheres-the-fire.
Sante, Luc. "Disco Dreams." *New York Review of Books*, 13 May 2004, 22–24.
Stuever, Hank. "Unspooled." Posted 29 October 2002, accessed 1 February 2006, http://www.hankstuever.com/death.html.
Tripper, Jack. "How to Make the Perfect Mix Tape." Tiny Mix Tapes. Posted 19 October 2001, accessed 23 July 2008, http://www.angelfire.com/indie/tinymixtapes/columns/10.15.01_how_to_make_the_perfect_mix_tape.htm.
Vinyard, Chris. "I'm Gonna Make You a Mixtape!" *Emmie Magazine*, 31 January 2006.
Wilson, Carl. "Cassette Mythos: Elegy." *Zoilus*, 4 June 2005. Accessed 1 February 2006 http://www.zoilus.com/documents/in_depth/2005/000455.php.

# 2

Constructing Discourses
and Communities

# Appeals to the Body in Eco-Rhetoric and Techno-Rhetoric

M. Jimmie Killingsworth

*Techno-rhetoric*—the study, practice, and teaching of electronic literacies, as in the fields of new media studies and computers and composition—may draw upon the same terminology as the rhetoric of place and environmental communication, or *eco-rhetoric*, but the aims of the two discourses still remain distinct. Such a point may seem obvious until you read the literature on both sides. A title such as Richard Selfe's *Sustainable Computer Environments* borrows not only the concept of *environment* but also that of *sustainability* from the environmental protection movement and thus hints at a close connection of environmentalist politics with techno-rhetoric, a field that might otherwise seem accommodationist in its promotion of corporate technology, at least from the viewpoint of the anti-corporate environmentalist. But accommodationist murmurs also arise on the green side of the exchange. Sid Dobrin, for example, in his eco-compositionist manifesto "Writing Takes Place," explicitly claims that eco-rhetoric should not stop at the study of geographical sites but should also include the presumed ecology of computer classrooms and Web-based environments. Clearly we are well beyond a simple dichotomy between luddite and cyborg rhetoric. But we may also be beyond the trend simply to deny the opposition, a trend that begins with Donna Haraway's eloquent pronouncements from the early nineties about "the leaky distinctions between animal-human (organism) and machine" (152; see also Mazlish).

Rather than a simplistic dichotomy between the discourses on the organic and mechanistic modes of life, or an equally simplistic conflation of the two, what experience often puts before us is a continuum, a systematic relationship that flows from earth to organism to machine and back again, the general outline of which is given in figure 4.1.

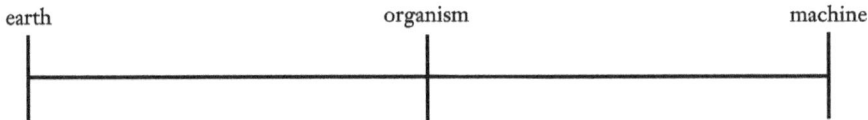

Fig. 4.1 The earth-organism-machine continuum

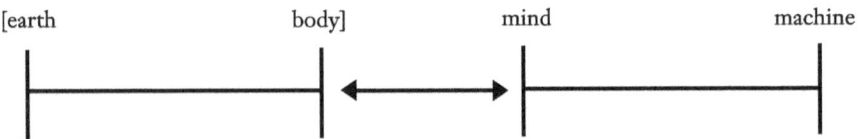

Fig. 4.2 Body divided from mind and bracketed with earth as inconsequential, as in techno-rhetoric

According to this view, the difference between eco- and techno-rhetoric frequently involves which part of the continuum one chooses for a focus—the earth-to-human or the human-to-machine connection. I believe that the merger of the two discourses, which might be warranted in light of their ultimate continuity, founders on the issue of the existential body. More specifically, techno-rhetoric ostensibly accepts the earth-organism-machine continuum, but tends to preserve the old Cartesian worldview that divides body from mind. It then strives transcendentally to negate the body and earth—or bracket them as inconsequential—then treat the organism purely as a mind in communion, or even identical, with the machine (fig. 4.2).

Two points need extra emphasis here. First, I am mainly talking about discourse, not organic and mechanistic life per se. I am not talking about the dangers of computer games, for example, but about the dangers of the ways we talk and think about computer games. Second, I am not suggesting a clear binary opposition between earth-oriented and machine-oriented ways of being (or ways of talking for that matter) and thus indulging in a naïve retreat to essentialism; indeed I would argue that the earth-organism-machine spectrum offers a clear instance in discourse studies of what Sharon Crowley calls "the postmodernists' restless resolution of dualisms into continua" (182). I will come down more strongly in favor of the integrity of eco-rhetoric because, despite its affinity with old romanticist models of discourse, eco-rhetoric is more likely to engage the full length of the continuum. By contrast, techno-rhetoric, in spite of its greater likelihood to claim an affinity with postmodernism, too frequently turns out to be some version of Cartesian modernism in a terminological masquerade, weakly appealing to a posthumanist paradigm, environmental awareness, and embodiment.

I begin with a story from my own techno-autobiography that is an allegory of the contemporary neglect of the body in techno-rhetoric. The suffering body becomes the phenomenological focus that resists the smooth substitution of virtual (machine) worlds for the physical (earthly) world. The machine may continue to run as long as there is fuel, but the bodily interface fails in a way that anticipates a more general atrophy or collapse, the sapping of the earth and the overuse of energy resources. Next I extend the critique to the texts of techno-rhetoric, specifically to the literature on computers and writing, to expose the neglect of the body in the relentless promotion of technological approaches to literacy. On the way to demonstrating the ultimate attempt at erasure of the body in techno-rhetoric and the contrasting recovery of the body in eco-rhetoric, I focus in particular on the concepts of extension and prosthesis as a way of conceptualizing technology's relation to the body and the earth. In this analysis, eco-rhetoric proves more likely to give a full account of the earth-body-machine spectrum, albeit an account that often (but not always) rejects the technological imperatives of the modernist perspective. Concluding, I consider briefly some practical and theoretical consequences of the neglect of the body in techno-rhetoric. These considerations apply specifically to the research and teaching of writing programs in the American university.

### The Allegory of the Prosthetic Demigod

The story from my techno-autobiography betrays a surprising affinity with an episode from the television program *South Park* called "Make Love, Not Warcraft," in which bodily health declines as technological competence advances. In the episode, the South Park kids suffer an outbreak of obesity and bad skin as the price they pay to become masters of an electronic game. The story of their neglect of the body stands as an allegory in the contemporary rhetoric of technology and human experience. Although different in the particulars, my story follows a similar symbolic pattern.[1]

In May 2006 I read a paper at the Computers and Writing Conference in Lubbock, Texas.[2] The point of the paper was to question use of the term "environment" in eco- and techno-rhetoric. The term long ago fell into disfavor among some eco-rhetoricians because it implies that nature is merely "that which surrounds," connoting a necessary separation of nature and humanity (Killingsworth and Palmer 42–44). Eco-rhetoricians tend to prefer a term like "lifeworld." Drawn from the philosophy of phenomenology, lifeworld suggests an intimate connection between organism and place. It connotes Heidegger's revision of "being" as "being there" (*Dasein*)—being as situated in the world. The term would be less amenable to techno-rhetoric, however, because it seems to favor the carbon-based world of organic life over the silicon-based experience of electronic devices.

We could leave it at that and say eco-rhetoric is concerned with the lifeworld, whereas techno-rhetoric really is more concerned with environments, the artificial surroundings of organic life. But in the spirit of continuity, the original paper sought out a term that both sides could embrace. The concept of "extension" seemed to suffice—the human body as an extension of the earth's body, and technology as an extension of the human. The organism extended in two directions thus becomes the mediating point in the continuity between technology and the earth (as suggested in fig. 4.1).

As we use the concept of extension today, the most immediate influence is the godfather of new media studies, Marshall McLuhan. His book *Understanding Media: The Extensions of Man*, first published in 1964, includes the essay "Clothing: Our Extended Skin." Clothing could be considered an environment that envelops the body, but generally it is too close and too portable—it clings to the body and goes with you everywhere (at least in public)—so it seems instead an extension, your public skin, a counterpart to your mental persona. McLuhan's own most obvious theoretical source is Sigmund Freud's 1930 book *Civilization and Its Discontents*. Freud understands the extensions of technology as an attempt to address people's anxieties over the inadequacies of the body. In earlier times people saw themselves, in Freud's words, as "feeble animal organism[s]." They "formed an ideal conception of omnipotence and omniscience" that, being denied to humanity, could be embodied in the gods. "Today," says Freud, "[the human being] has come very close to the attainment of this ideal"—becoming "almost . . . a god." The new god builds "auxiliary organs" to extend the body's powers—microscopes and telescopes to extend vision, communication devices to send the voice around the globe, airplanes to fly, clothing, armor, and then fortresses to add layers that protect the tender skin. When this "prosthetic God" dons all these "auxiliary organs," in Freud's view, the result is "truly magnificent" (Freud 44–45).

This version of extension, the concept of the prosthetic, has proved appealing in techno-rhetoric, where it has been politicized to some degree. In "Wearable Computing as a Means of Personal Empowerment," Steve Mann suggests that if prosthetics are used to replace missing limbs, to take them away from the wearer would be a violation of human rights. Applying the term "prosthetic" to technological devices is, in effect, to endow them with the same set of rights. The wearer is entitled to have regular access, for example, to "wearable memory."[3] By extension, to take away cell phones from students in class amounts to denying their connection to the world.

From McLuhan and Freud and Steve Mann, it is an easy step to imagining the contemporary, computer-enhanced professor as a version of the prosthetic demigod. Armed with my technological extensions, I can sit at a home computer and do what used to require a far greater expenditure of time and effort. I can write faster and more accurately than I could with a pencil or a typewriter. I can do research by consulting online databases instead of going to the library. I can teach and grade papers without going to the classroom. I can confer with students and colleagues without going to the office. I can attend conferences without getting on the airplane. My university administration loves me for saving travel money and classroom space. The librarians love me because they can realize their venerable but necessarily unstated ideal of keeping every book on the shelf and out of use at all times. My students love me because they do not have to get up at eight o'clock, or ten, or even noon to meet with me or come to class.

Of course, there is the nagging fear of a system failure or electrical outage that would severely diminish if not totally disable my extended power. Freud anticipated such worries and was not willing to stop with a happy image of power and productivity. At the time he wrote his famous portrait of the prosthetic god, he was suffering from mouth cancer and was forced to wear an ill-fitting prosthetic jaw, so his awareness of the shortcomings of technology was all the more acute. He pointed out that the "organs" of the "prosthetic God" "have not grown on to him and . . . still give him much trouble at times." Although taking some comfort from the thought that things might get better in the future, Freud concludes, "present-day man does not feel happy in his Godlike character" (44–45).

Less frequently discussed than the possibility of a bad fit between prosthetic and human being is the failure of technology on the human side. Which brings me back to my story. After months of testing my ideas on techno-rhetoric by the fullest possible immersion in the technologies of writing—in wikis, blogs, Web sites, and word processing—I barely made it to Lubbock to read my paper. I had decided to drive across Texas for the meeting, but by the time I actually undertook the trip, I was having a terrible pain that, as I found out later, came from two herniated disks in my neck and upper back. The trip to Lubbock was a physical challenge. Once there I could barely sit in a chair long enough to get through a conference session. I could walk for miles around the lovely campus of Texas Tech, where I saw scissor-tailed flycatchers cutting patterns in blue sky and jackrabbits grazing like cattle. But sitting was a huge problem. I had reached my limits as a prosthetic god.

My body had become this uncooperative *thing*, this *other* that resisted my technological ambitions. I took it to the shop—the doctor, the chiropractor, the physical therapist. And after a year of therapy, I had avoided surgery and brought body and soul back into rough harmony again. Part of the price, however, was to limit computer use and time spent in the sitting posture. I revived my use of the notebook—the carbon-, not the silicon-based notebook—and I spent more time outdoors and in the gym. My research in eco-rhetoric continued to flourish, while my work in techno-rhetoric languished.

In this condition, I offer myself as an allegory. Like the business that overextends financially by opening too many stores, or the army that overextends its lines of communication and thus opens itself to flanking maneuvers, I had overextended my body, favoring certain postures (sitting), certain behaviors (reading screens), and certain senses (especially sight) while neglecting others (walking, standing, listening). A singer may overuse the voice even with the aid of a microphone; driving the car too much is bad for the back. Our extensions still connect to the body and stress it in very particular ways. Such matters are clichés in the preventive health and physical therapy business.

But the allegory of the prosthetic demigod points to a further truth: What is happening to the body is happening to the earth on a larger scale. The idea is well known in the field of eco-rhetoric. The mother of modern environmentalism, Rachel Carson, made the point explicitly in her influential exposé of the pesticide industry, *Silent Spring*, which she wrote in the early 1960s as she was dying of a cancer that could well have been caused by environmental influences. "There is an ecology of the world within our bodies," she wrote; like all of organic nature, we trade in "the common currency of energy" (170, 185).

### Embodiment in Techno-Rhetoric

In techno-rhetoric, the same dissatisfaction that feeds the consumerist culture of fad diets, plastic surgery, and personal trainers drives the interest in enhancement and extension. For Steve Mann, as for Freud, all bodies are disabled. Prosthesis is not the exclusive practice of the blind, the elderly, the physically handicapped; we are all naturally disabled. Small, feeble, vulnerable, our bodies constantly victimize us, frustrate us, deny our ambitions. Poor health is not a sign of dysfunction or trouble, as it is in eco-rhetoric; it is the norm of the human condition.

This attitude comes through clearly in science fiction, one of the fountainheads of techno-rhetoric, and no influence from this quarter is greater than William Gibson's 1984 cyberpunk novel *Neuromancer*, written on a Smith-Corona typewriter ten years before the Internet took hold of public communication. Gibson envisages a world where technology allows its most competent adherents to live in a heaven of light and power, a matrix of pure mentality, a gee-whiz realization of the old Cartesian dream of body-mind dualism, in which the mind proves transcendent and outstrips the limits of what Gibson's characters call "the meat world." For the book's protagonist, the cybercowboy called Case, the "matrix" or "cyberspace" stands for freedom, whereas the body is viewed as a death trap, the mind's prison. The ironic narrative voice of the novel, although intrigued with the possibilities Case's behavior suggests, skeptically probes his attitude toward the mind/body complex. The narrative returns attention to the needs of the body again and again, pleading the case of interpersonal involvement against self-absorbed addiction. In final analysis, Gibson portrays Case as an "artiste" certainly, but above all as a *case*, as in "case study" or "mental case." In the years following *Neuromancer*, adherents of the Web borrowed the term "cyberspace" for the product they developed and promoted, and the producers of the film *The Matrix* took the synonymous term to represent a hell of misperception foisted on humanity by increasing dependence upon, and finally defeat by, the machines, which ultimately enslave the collective body of humanity while treating the mind to a pleasant consensual hallucination, to use the terms of Gibson's novel.

If we extend this interpretive thread only a bit further, using the terminology introduced by the geographer Yi-Fu Tuan, we can say that the body stands for place, whereas the matrix represents space. Stating both terms positively, Tuan writes, "Place is security, space is freedom, we are attached to the one and long for the other" (3). Places are endowed with flora and fauna, indigenous and introduced, with geography and a characteristic terrain, with people and their special cultures and history; space is open, waiting to be planned, a yet-to-be-cultivated field, an unpopulated expanse of the world yielding to the imagination of the person who can acquire it. The two terms often collide politically. In the nineteenth century, Euro-American settlers referred to the western lands as the wide-open spaces, but the indigenous peoples understood the same lands as places, hunting grounds, homes. In this sense, a human body is the place of places, always specific and characteristic of a person; only as an abstraction can we think of the body as a space. To turn the body into an other, a space or a thing, as happens in torture, war, injury, or disease, is, in the language of Elaine Scarry, to "unmake the world" that the person inhabits.

This theoretical connection of body with place allows us to see more precisely how eco-rhetoric departs from techno-rhetoric. If techno-rhetoric resents the demands of the body and seeks to remake it in the image of the machine, overcoming its limits with extensions and enhancements, eco-rhetoric celebrates the body's connection to the earth and strives to accept the limits of the body as part of the perpetual struggle against the human hubris and overreaching that deplete resources and erode the earth. In this sense, eco-rhetoric departs from the old humanist model of the mind-body-earth relationship—seeking to purvey an ecocentric or biocentric worldview rather than an anthropocentric one—but techno-rhetoric sticks with the dualistic model. As Katherine Hayles has argued, the so-called cybernetic posthuman shares with the old humanist self of Cartesian dualism and imperialist universalism at least one feature: what she calls "the erasure of embodiment." "Identified with the rational mind, the [humanist self] *possessed* a body but was not usually represented as *being* a body," says Hayles (4). Likewise, for the cyberhuman of the postmodern world, the body is not the core of identity so much as an element in a distributed identity that includes machines as well as other people. The problem of thus identifying the body with machines is that we may come to think of the body—and by extension, other people—as something we use. Becoming users of the body, rather than a body itself, we are prone to *over*use or even *ab*use the body.

Techno-rhetoricians such as James Paul Gee are inclined to admit that humans think best "through their bodies and emotions" in situated learning, but when Gee and his comrades actually come to discuss a favorite term of the new cognitivism, "embodiment," they seem concerned more with the mind than with the body itself ("Foreword" x). In their world, the mind is embodied not only in the carbon-based shell of earthly existence but also in the electronic body of the Web surfer or the gaming avatar.[4] This robotic body allows the mind the freedom to roam a worldwide shopping mall or kill boars in a medieval forest, achieving levels of satisfaction, competence, and control unknown in carbon-based life—all while the participant never moves from the sitting posture, eyes fixed on a screen of dazzling imagery.

The experience has a distinct, if furtive, eroticism. Debra Journet thus writes of being "seduced" by the game of Myst, captivated by the "beauty" of its enticing "landscapes," obsessed with spending time in the virtual world (97, 103). Indeed, the language of obsession, compulsion, and addiction—states of mind usually considered harmful to bodily health—haunts the literature of techno-rhetoric. On the much-discussed topic of "cybersex," Sherry Turkle writes, "An Internet list of 'Frequently Asked Questions' describes . . . cybersex . . . as people typing messages with erotic content to each other, 'sometimes with one hand on the keyset, sometimes with two.'" True to the treatment of the mental bias in this literature, Turkle reminds us of "the adage that ninety percent of sex takes place in the mind" (21). But what interests me is the furtiveness of that hand missing from the keyboard. It seems to disappear from the scope of the investigation in a verbal gesture at once prudish and titillating. The body—in this case, the actual genitalia toward which the hand reaches—is outside the scope, the techno-rhetorical line of vision. More frequently even such oblique references disappear in techno-rhetoric, and only the language of sexual attraction remains, as in the essay by Journet.

## Bodily Involvement in Eco-Rhetoric

By contrast, eco-rhetoric favors a complete identification of person with body. I do not use a body; I am a body. And I am part of a world that is not an extension of my desires and fantasies, not a space that I possess, but rather a home place out of which my body grows, the health of one relying upon the health of the other. Abandoning even the extension of clothing as an extended skin, eco-rhetoricians favor nakedness with their sensually varied and particularly tactile imagery (as opposed to the scopophilic and obsessive dependency on the gaze in the visual rhetoric of pornography, the very lifeblood of the Internet). Likewise valued in eco-rhetoric is unassisted or minimally enhanced physical power. It is said that you can define your bioregion, your home place, by how much ground you can cover on your own power in one day, walking or at most biking and canoeing.

In a work often cited as an early instance of ecocritical nature writing, the 1882 memoir *Specimen Days*, the poet Walt Whitman returned to a theme that first attracted him in the earliest edition of his masterwork, the 1855 *Leaves of*

*Grass*, in which he avowed, "I will go to the bank by the wood and become undisguised and naked, / I am mad for it to be in contact with me" (13). Writing as an old man, after he had survived a paralytic stroke, Whitman describes his experience of nature this way: "It seems as if peace and nutriment from heaven subtly filter into me as I slowly hobble down these country lanes and across fields, in the good air, as I sit here in solitude with Nature—open, voiceless, mystic, far removed, yet palpable, eloquent Nature. I merge myself in the scene, the perfect day" (806). The paradoxically "voiceless" yet "eloquent" earth affirms identity not by way of language and the mind, but through the senses and the body of the old poet, who says, "Somehow I seem'd to get identity with each and every thing around me. . . . Nature was naked, and I was also" (807). To become a body among bodies, a flow among flows, to let go of a defining vision and a categorizing language, the means by which the human mind is extended in communication, is to experience the fullness of naked contact with the earth, in Whitman's mystical view. (For a further discussion, see Killingsworth, *Walt* chap. 6.)

For the romantic poet, as for the eco-rhetorician, language itself can prove problematic. To refer to the earth as naked, for example, involves a metaphorical imposition. Nakedness implies a contrast to the state of dress that is characteristically human. People get naked; they go au naturel; but nature cannot, at least not literally. The poet, undressing himself, feels an identity with the earth, expressed in metaphor, the trope of identity. Earth is as it was before, but he declares it naked.

Metaphor, like clothes and the computer, is always an extension, always prosthetic. As I. A. Richards suggests, metaphor transfers terminology from one context to another. With frequent use, the original context fades from consciousness. It ceases to produce an informing tension and becomes instead a vague and often troubling resonance. We might forget, for example, that when we speak of visiting a Web site, we are using a metaphor. But at the edge of awareness is the realization that visiting an online shopping site is very different from visiting a neighborhood store, although the effect on the local economy of sales lost to Internet sources might be very real indeed. Again notice the political conflict between space and place—place as security, space as possibility—this time realized through the power of metaphorical rhetoric.

In *Desert Solitaire*, Edward Abbey studiously resists the extensional powers of language, most memorably in a passage that stands as a manifesto against personification (see Buell 180–218). A balanced rock appears to him in the red Utah desert as "a stone god or a petrified ogre," but then he draws back from the comparison:

> Like a god, like an ogre? The personification of the natural is exactly the tendency I wish to suppress in myself, to eliminate for good. I am here not only to evade for a while the clamor and filth and confusion

of the cultural apparatus but also to confront, immediately and directly if it's possible, the bare bones of existence, the elemental and fundamental, the bedrock which sustains us. I want to be able to look at and into a juniper tree, a piece of quartz, a vulture, a spider, and see it as it is in itself, devoid of all humanly ascribed qualities, anti-Kantian, even the categories of scientific description. . . . I dream of a hard and brutal mysticism in which the naked self merges with a non-human world and yet somehow survives still intact, individual, separate. (7)

Abbey's "bedrock" is not rock, however, nor the immediate ever immediate (that is, free of mediation), because at the base of existence for human perception is always the body, the first medium that defies immediacy and the ground of every perception. Abbey may lay claim to a desire to see earth-as-it-is but cannot resist the language that transforms the earth into a great body—"naked" but still metaphorically kin to the naked human body that greets it—or something recognizable as existing at the edges of bodily life, a corpse with its "bare bones."

Abbey's resistance to troping is a study in the way language returns ever to metaphors of the body, as Lakoff and Johnson have shown in their tour de force *Metaphors We Live By*. No matter how elaborate a metaphor becomes, its ultimate point of reference is the body, and more specifically the relation of the body to the earth. The best example is the conceptual constellation involved in the word "depression." "I'm depressed," I say, or "I'm down," a literal symptom of which is that I cannot get out of bed in the morning. I cannot get up, arise. If I get low enough, suicide looms; gravity draws me to the grave—the gravest conclusion of depression (see Killingsworth, *Appeals* chap. 9).

Abbey's struggle with personification parallels his resistance to technology—resistance that differs from the techno-rhetorical denial of the body in that it does not involve forgetfulness. Indeed it is crucial to this discourse to stay mindful of the entire spectrum of human experience. At one point in his memoir, Abbey offers a self-disparaging account of composing a letter in his ranger's trailer under light extended beyond the daytime with the help of an old generator that "sputters, gasps, catches fire, gains momentum, winds up into a roar," and finally settles into an obnoxious whine (15). The mechanical thing, portrayed here in metaphors of a sick or broken body with its gasping and whining, or a dragon with its fire and roar, produces an unnaturally bright light that blinds him before it enables him to settle down to writing in an ambivalence of extended ability gained at the cost of a ruined peace of mind. In questioning the use of a flashlight when he walks in the desert at night, to take a milder example, he concedes that it is a "useful instrument" but insists that "I can see the road well enough without it. Better, in fact" (14–15). And there is the larger problem: "like many mechanical gadgets it tends to separate a man from the world around him. If I switch it on my eyes adapt to it and I can only see the small pool of light which it makes in front of me; I am isolated. Leaving the flashlight in my pocket where it belongs, I remain a part of the environment I walk through and my vision though limited has no sharp or definite boundary" (15). In accepting the limits of the body's power, Abbey finds it more to his liking than what most of us would consider a technological extension of its power. The flashlight's capacity to light up the night and dispel the perennial fear of the dark ultimately limits natural ability.

Even though Abbey's resistance to metaphor and technology proves impossible to sustain, he compensates by cultivating mindfulness of limits, scope, and range, the very characteristics that techno-rhetoric seems most eager to ignore, outrun, or overcome. Part of reclaiming a sense of place for Abbey involves reclaiming the bodily awareness numbed by technological experience. He contends that "you can't see anything from a car; you've got to get out of the goddamned contraption and walk, better yet crawl, on hands and knees, over the sandstone and through the thornbush and cactus. When traces of blood begin to mark your trail you'll see something, maybe" (xii).

We might be inclined to dismiss the blood-and-bones outlook of old Abbey as an instance of an outdated literary machismo. But it has proved surprisingly sustainable in the discourse of naturalism, in women's writing as well as in men's. Annie Dillard begins her now-classic book of 1974, *Pilgrim at Tinker Creek*, with the story of her old fighting tomcat coming in through the open window of her bedroom and landing on her chest in the night. "I'd wake up in daylight to find my body covered with paw prints in blood; I looked as though I'd been painted with roses," she writes: "The signs on my body could have been an emblem or a stain, the keys to the kingdom or the mark of Cain. I never knew" (9). Dillard's parable for the writer in the first chapter of *The Writing Life* is the story of an Algonquin woman who, instead of starving to death in an arctic winter, uses a strip of flesh from her own thigh to catch a fish under the ice and thus save herself and her infant. "The materiality of a writer's life cannot be exaggerated," she insists (576). "The art must enter the body" (590). In this same vein, Dillard's younger sister in nature writing, Janisse Ray, tells us in *Ecology of a Cracker Childhood* that the memory of the lost pine forests of her south Georgia home "is scrawled on my bones, so that I carry the landscape inside like an ache" (4). And as noted earlier, Rachel Carson provided a scientific foundation for the blood-and-bones school of nature writing.[5]

We can thus say that on the earth-organism-machine continuum, eco-rhetoric struggles in one direction and resists in the other. It struggles to regain the connection of organism to earth that technology inhibits and that language can overpower, reversing the dependency of humanity on its earthly sources. But to resist is not to deny or forget, and that is the key difference between eco- and techno-rhetoric. I am not saying that eco-rhetoric is always more successful than techno-rhetoric, or necessarily more advanced or enlightened; only that in current manifestations it is more likely to keep all the elements on the continuum in play.

### Consequences—Practical and Theoretical

Now we come to the big question that every rhetorical critic rightly fears: so what? So what if techno-rhetoric breeds forgetfulness of geographic places and neglects the experience of bodily presence? Nature writing is boring after all, and we are mainly an urban people. And so what if the body of some *gringo viejo* in an oversized Texas university becomes, to borrow an image from Greg Brown's song "Slant-Six Mind," roadkill on the information highway? What are the real consequences?

First and foremost is the tendency to forget about the demand of silicon-based writing and teaching on the energy supply. A discourse of forgetfulness diminishes awareness of the electrical uptake required to make thousands of computers run all day and all night in most every house and office around the country. A nice clean connection to a virtual world usually depends upon a much dirtier connection to a coal-fired power plant somewhere near somebody's home place. I have never read an environmental impact statement as part of a plan to install a computer classroom or to increase the use of computers in a writing program. Indeed, it is difficult to find studies of how much energy computers actually use, even with the easy access afforded by Web-searching engines and such databases as the online *Applied Science and Technology Index*. Searching under rubrics like "computer energy use" and "environmental impact, computers," what you do find is a large number of articles telling you how to save on energy costs by shutting down your computer at night (a little high-tech laptop can use as much electricity as an old-tech refrigerator, I learned) and how difficult it is to recycle computer parts. The images of landfills teeming with plastics and metals from discarded computers should raise big questions about software giants who render our equipment obsolete with every new version of their products. Should we really need to replace faculty workstations every three years, as the current wisdom at my institution suggests (an uncanny parallel with the shelf life of textbooks in a comparable industry)? The logic is that it is cheaper to replace than to repair after a certain number of years—a logic again driven by the availability of parts in an industry that keeps the "new and improved" and "more powerful" models coming out every year (often with functionalities that most users never learn to use before yet newer models appear). What is the environmental cost of such planned obsolescence? Should not this discussion engage scholars in techno-composition as much as the concern with "environmental footprints" has engaged such scholars as Derek Owens in eco-composition?

A second consequence, one that should worry directors of writing programs and department chairs, is the neglect not just of *the body* but of bodies. In my university, proposals for new hardware almost never fail, and proposals for software

are only slightly less successful. Where we run into problems is with proposals for "meatware," new technical staff and new teachers to make the machines run. The problem, I guess, is that you can throw old machines on the ever-larger scrap heap, but you have to take care of people, and that is an expensive business. As Reilly and Williams argue, questions of technology are always entwined with the politics of labor. In the literature on techno-rhetoric, there is an increasing interest in the question of labor, no doubt, but not enough to ensure the level of body awareness I have in mind nor to prevent the kind of habitual denial or neglect that I stand against. The few pages on "Bodies at Work" in Rob Shields' *The Virtual*, for example, are mostly devoted to showing that, despite public worries, no clear evidence connects carpal tunnel syndrome to excessive keyboard use (147–50).

Finally, a more subtle consequence, one more in line with the methods of analysis in this essay, concerns research and theoretical issues in the fields of literacy studies, English composition, rhetoric, and literary criticism. Habitual neglect of the body left seated at the computer (or wallowing around with the newer user interfaces) can lead to a willful blindness that spreads outward from the individual to include issues conspicuously related to bodily experience in social contexts, such as gender, class, and race, as well as the themes of hate, war, and violence (see Killingsworth, *Appeals* chaps. 6, 7, and 8; also Crowley). In the concern with access to technology and the identity issues covered in collections such as those of Selfe and Hawisher, techno-rhetoric may seem to be covering this crucial connection of body with identity politics, but without a clear account of the material foundation of such problems, political insensitivity and quietism can slip in the back door. Among the most prominent promoters of technology as a key to improving literacy, James Paul Gee is perhaps the most culpable in this regard. In his first book extolling the educational virtues of electronic gaming, he notoriously dismisses questions of gender and violence in explaining the uneven appeal of first-person shooter games. "I have nothing whatsoever to say about these issues," he writes (*What Video Games* 10). But before leaving the topic, he does manage to assert that "the issue of violence is widely overblown" and that "shooting is an easy form of social interaction (!) to program"—glibly adding an exclamation point in parentheses after the phrase "easy form of social interaction" to register what is, I suppose, some measure of shock at himself for being able to write such a thing. In his rather defensive foreword to Selfe and Hawisher's *Gaming Lives in the Twenty-First Century*, Gee moves on to race. "I do not think that the issue of race and games is just that some games are racist," he writes. "They are no more or less so than the U.S. media culture they give back to us" (xii). We might just as well excuse a student paper full of hate language and verbal abuse as no worse than the homophobic, misogynistic, and racist culture at large. But we do not excuse such writing. We call the school psychologist or the campus police.

The problem is that Gee either has no understanding of the critical function of rhetorical analysis (not very likely), or in an act of willful avoidance, he turns off the switch so that he can get at what is good in video games without worrying about what is bad. Along with some other writers on electronic literacy (such as DiSessa xi), Gee insists on this practice of cultivated critical incapacity, avoiding the "negative" in order to better comprehend the "positive." However, by dismissing questions of gender, violence, and race from his considerations, and by more generally neglecting the place of the body in the earth-organism-machine continuity, Gee raises doubts about his entire project. He adamantly insists, for example, that the enhanced learning ability of home gamers accounts for their success in school. But how would he know? Could such success possibly arise not from enhanced literacy but from the cultivation of a special will to power, a killing competitiveness fed by the control fantasies in a steady diet of graphic violence and the pursuit of a superexpert competence in a narrow range of highly specialized skills? Or could it be that the kind of literacy Gee values is one that powerfully concentrates the attention to a limited scope while just as powerfully crippling awareness of whole other fields of experience? As rhetoricians we cannot ignore how even the most remote connections among the elements of experience—bodies, machines, social structures, attitudes, fantasies, ideals—are formed and reinforced. We want instead to ferret out forgotten sources and bring hidden assumptions to light.

In this cultivated critical incapacity, Gee may not be alone in our field. Sharon Crowley admonishes that rhetoric and composition as practiced in English departments lag behind the discipline of speech communication in the development of rhetorical criticism. Indeed she says that it is "virtually absent from composition studies" (185). Stuart Selber is moving in the right direction when he insists on including "critical literacy" in his concept of a multiliterate world that treats technological literacy, or competence, alongside print literacy. I would also want to add what David Orr years ago called ecological literacy. And more to the point of this paper, I would again stress Crowley's main point—that body criticism offers a path into fuller realization of critical rhetoric.[6]

As for critical techno-rhetoric, another promising possibility would involve a fuller treatment of the erotics of technology, a topic that hovers on the edges of my own analysis as well as that of such authors as Sherry Turkle and Debra Journet. Like rhetoric, the erotics of technology will ask, What is the appeal? What holds the attention? What moves and pleases us? Again, a good starting place for such a study is to confront questions about not only the social body but also the bodies of individual users, the ones addressed in invitations to porn sites, electronic shopping venues, and dating services online, from the *gringo viejo* forced to delete hundreds of ads every year that promise help for erectile dysfunction to the lonely lad enticed by the possibility of realizing his most disturbing fantasies. What is the appeal? What holds the attention? What moves us? And why?

## Notes

1. Thanks go to David Cockley of Texas A&M for calling my attention to this episode. I would also like to thank Elizabeth Talafuse for finding the episode for me on YouTube and helping in other ways with the research for this paper. Further help came from Sarah Hart.

2. Parts of the original paper are reproduced in summary in this section. Parts also appear in Killingsworth, "A Phenomenological Perspective."

3. Thanks go to Professor Isabel Pedersen of Ryerson University in Toronto for calling my attention to the work of Steve Mann.

4. Sharon Crowley quotes Halbertstam and Livingston in her brief overview of the "posthuman body," which is said to be "a technology, a screen, a projected image" that "both writes and is written upon" (Crowley 178). I intentionally address the concept of the posthuman only tangentially here because, although I admire much about the treatment of the body by such writers as Hayles and Crowley, who seem content with the terminology of posthumanism, I have a problem with the slippage of metaphors like the one quoted from Halberstam and Livingston toward the literal treatment of "screened" bodies in writers such as Gee as a replacement for the blood-and-bones body of a more existential or phenomenological reading of physical life. The trouble is that some applicants of the posthumanist terminology tend to miss the crucial ironies of a Katherine Hayles or Donna Haraway.

5. For more on the rhetoric of Rachel Carson, see Killingsworth and Palmer 64–68; also Waddell. For a somewhat fuller treatment of Janisse Ray's appeals to place and the body, see Killingsworth, *Appeals* 63–66.

6. See Jack Selzer's groundbreaking collection on body rhetoric, in which Crowley is a contributing editor; also Debra Hawhee's reconsideration of the body in classical rhetoric.

## Works Cited

Abbey, Edward. *Desert Solitaire: A Season in the Wilderness.* New York: Ballantine, 1968.
Buell, Lawrence. *The Environmental Imagination: Thoreau, Nature Writing, and the Formation of American Culture.* Cambridge, Mass.: Harvard University Press, 1995.
Carson, Rachel. *Silent Spring.* New York: Fawcett Crest, 1962.
Crowley, Sharon. "Body Studies in Rhetoric and Composition." In *Rhetoric and Composition as Intellectual Work,* edited by Gary Olson, 177–87. Carbondale: Southern Illinois University Press, 2002.
Dillard, Annie. *Three by Annie Dillard: Pilgrim at Tinker Creek, An American Childhood, The Writing Life.* New York: Harper, 1990.
DiSessa, Andrea A. *Changing Minds: Computers, Learning, and Literacy.* Cambridge, Mass.: MIT Press, 2001.
Dobrin, Sidney I. "Writing Takes Place." In *Ecocomposition: Theoretical and Pedagogical Approaches,* edited by Christian R. Weisser and Sidney I. Dobrin, 11–25. Albany: State University Press of New York, 2001.
Freud, Sigmund. *Civilization and Its Discontents.* Trans. James Strachey. New York: Norton, 1989.
Gee, James Paul. "Foreword." In Selfe and Hawisher, *Gaming Lives,* ix–xii.

———. *What Video Games Have to Teach Us about Learning and Literacy*. New York: Palgrave, 2004.
Gibson, William. *Neuromancer.* New York: Ace, 1984.
Halberstam, Judith, and Ira Livingston, eds. *Posthuman Bodies*. Bloomington: Indiana University Press, 1995.
Haraway, Donna Jeanne. *Simians, Cyborgs, and Women: The Reinvention of Nature*. New York: Routledge, 1991.
Hawhee, Debra. *Bodily Arts: Rhetoric and Athletics in Ancient Greece*. Austin: University of Texas Press, 2004.
Hayles, N. Katherine. *How We Became Posthuman: Virtual Bodies in Cybernetics, Literature, and Informatics*. Chicago: University of Chicago Press, 1999.
Heidegger, Martin. *Being and Time*. Trans. Joan Stambaugh. Albany: State University of New York Press, 1996.
Journet, Debra. "Narrative, Action, and Learning: The Stories of *Myst*." In Selfe and Hawisher, *Gaming Lives*, 93–120.
Killingsworth, M. Jimmie. "A Phenomenological Perspective on Ethical Duty in Environmental Communication: A Response to Cox." *Environmental Communication* 1 (2007): 58–63.
———. *Appeals in Modern Rhetoric: An Ordinary Language Approach*. Carbondale: Southern Illinois University Press, 2005.
———. *Walt Whitman and the Earth: A Study in Ecopoetics*. Iowa City: University of Iowa Press, 2004.
Killingsworth, M. Jimmie, and Jacqueline S. Palmer. *Ecospeak: Rhetoric and Environmental Politics in America*. Carbondale: Southern Illinois University Press, 1992.
Lakoff, George, and Mark Johnson. *Metaphors We Live By*. Chicago: University of Chicago Press, 1980.
Mann, Steve. "Wearable Computing as a Means for Personal Empowerment." Keynote Address, 1998 International Conference on Wearable Computing, Fairfax, Va., 17 June 2003, http://wearcam.org/icwckeynote.html.
Mazlish, Bruce. *The Fourth Discontinuity: The Co-Evolution of Humans and Machines*. New Haven, Conn.: Yale University Press, 1995.
McLuhan, Marshall. *Understanding Media: The Extensions of Man*. Critical Edition. Edited by Terrence Gordon. 1964. Corte Madera, Calif.: Gingko, 2003.
Orr, David W. *Ecological Literacy: Education and Transition to a Postmodern World*. Albany: State University of New York Press, 1992.
Owens, Derek. *Composition and Sustainability: Teaching for a Threatened Generation*. Urbana, Ill.: National Council of Teachers of English, 2001.
Ray, Janisse. *Ecology of a Cracker Childhood*. Minneapolis, Minn.: Milkweed, 1999.
Reilly, Colleen A., and Joseph John Williams. "The Price of Free Software: Labor, Ethics, and Context in Distance Education." *Computers and Composition* 23 (2006): 68–90.
Richards, I. A. *The Philosophy of Rhetoric*. New York: Oxford University Press, 1936.
Scarry, Elaine. *The Body in Pain: The Making and Unmaking of the World*. New York: Oxford University Press, 1985.
Selber, Stuart A. *Multiliteracies for a Digital Age*. Carbondale: Southern Illinois University Press, 2004.

Selfe, Cynthia L., and Gail E. Hawisher, eds. *Gaming Lives in the Twenty-First Century: Literate Connections.* New York: Palgrave, 2007.

———. *Literate Lives in the Information Age: Narratives of Literacy from the United States.* Mahwah, N.J.: Erlbaum, 2004.

Selfe, Richard. *Sustainable Computer Environments: Cultures of Support in English Studies and Language Arts.* Cresskill, N.J.: Hampton, 2004.

Selzer, Jack, and Sharon Crowley, eds. *Rhetorical Bodies.* Madison: University of Wisconsin Press, 1999.

Shields, Rob. *The Virtual.* London: Routledge, 2003.

Tuan, Yi-Fu. *Space and Place: The Perspective of Experience.* Minneapolis: University of Minnesota Press, 1977.

Turkle, Sherry. *Life on the Screen: Identity in the Age of the Internet.* New York: Simon and Schuster, 1995.

Waddell, Craig, ed. *"And No Birds Sing": The Rhetoric of Rachel Carson.* Carbondale: Southern Illinois University Press, 2000.

Whitman, Walt. *Complete Poetry and Collected Prose.* New York: Library of America, 1982.

# Unfitting Beauties of Transducing Bodies

Anne Frances Wysocki

It matters, of course, the understanding of "persuasion" one has in mind while discovering the available means thereof—and my understanding in this essay depends on the last roughly half-century of attentions to means of shaping behavior and identity that are non-linguistic and that appeal, usually quietly and without direct address, to bodies and feelings rather than articulated logics. It is persuasion that follows not from a decision made inside one's mind but rather from a sinew or pulse shifting, and perhaps staying shifted, in response to something meant to shift it. I could point here to recent rhetorical analyses of specific spaces—such as pulpits, battlefields, or a Starbucks store—that explain how each space is a "physical representation of relationships and ideas" and so encourages those moving within toward particular attitudes and relationships (Mountford 42; see also, for example, Halloran; Dickinson; Blair, Jeppson, and Pucci; Fleming). My focus is instead on experiences meant to shape our senses of our selves by shaping our senses themselves, such as when Debra Hawhee considers the "network of educational and cultural practices articulated through and by the body" in the ancient Greek overlapping of rhetorical and athletic training (6).

Here I look to more recent practices and technologies. How might some digital texts—some presented as art—impel us toward particular sensuous engagements with the world and each other? What are possible theoretical takes on those engagements—and what are implications of those takes? These questions are worth asking, I think, because—given recent shifts in technologies of production, distribution, and consumption—the texts we and the people in our classes consume but perhaps also produce (documentary videos, instructional Web pages, games meant to persuade us to become soldiers) can make questions of aesthetics more present than research papers traditionally have. When typewriters and college-ruled paper engaged our hands, we might have discussed (for example) the ways and roles of emotional appeals in academic writing; if we ask people in our classes to produce Web pages or to weave photography or even simple typography together with words, questions of how color, shape, movement on a page, or visual representation appeal become unavoidable. We could consider such appeals aesthetic in a loose sense—"Is that arrangement of colors pretty?"—but the concern here is a focused, temporally specific notion of aesthetics. This is aesthetics as a perspective for discussing embodied, sensuous responses to objects (including texts), for determining how and why some objects encourage us to judge them beautiful or otherwise; this is a perspective that became strong starting in the eighteenth century and that made aesthetic concerns inseparable from the ethical. In this essay I argue that, although we may want to hold a connection between the aesthetic and the ethical, we cannot if we act as though our bodies still fit eighteenth-century understandings of perception. By highlighting current aesthetic possibilities of our texts—digital as well as nondigital—we might practice having bodies that can alertly convert sensuous experience into ethical practice.

The art discussed here is less amenable to photographic representation than most painting or sculpture and so requires that I start with several long descriptive quotations. The quotations are meant to entice you, as an indication of the art's persuasively sensuous pull even in description; that pull—with its direction toward internal pleasure or toward external connection—motivates the arguments that follow concerning current theories about some new digital art.

Sabrina Raaf, a Chicago artist, produced the artwork *Saturday* in 2002. In her artist statement about *Saturday*, Raaf describes how she

> used walkie talkies, CB radios, and various other forms of consumer spy (or "security") technology in order to actively harvest [wireless] communication leaks. *Saturday* forms . . . [an] intimate portrait of the community of Humboldt Park, Chicago through a composite presentation of conversations stolen on Saturdays in the park. . . .
>
> The transmissions included communications between gang members on street corners nearby and group conversations between friends talking about changes in the neighborhood and their families. There were raw, intimate conversations and often even late night sex talk between potential lovers. . . . During the series of Saturdays, I also recorded the sounds of my neighborhood. . . . These are the sounds that are mixed in the piece. And these are the sounds that literally drip from participant's fingertips in *Saturday*.
>
> *Saturday* is presented in the form of an interactive glove. In order to hear the audio, participants *magically* just press their fingertips to their forehead and they hear the sound without the use of their ears. The glove is outfitted with leading edge audio electronic devices called

"bone transducers" which make this possible. These transducers transmit sound in a very unusual fashion. They translate sound into vibration patterns which resonate through bone. This is the same process as the natural hammer and anvil system inside our inner ears which allows us to perceive sound. Since the bone transducer does all this work artificially, it allows you to hear crisp audio without it being played out loud or through headphones. So, even if a user covered their ears and then placed their fingers to their temples, they still "hear" the sound.

This piece permits a new way of listening. The user places their fingers to their forehead—in a gesture akin to Rodin's "The Thinker" or of a clairvoyant—in order to tap into the lives of strangers. Pressing different combinations of fingers to the temple yields plural viewpoints and group conversations. These sounds are literally mixed in the bones of the listener.[1]

Another digital art piece, *Osmose*, was conceived by Char Davies and first exhibited in 1995.[2] A participant engages with *Osmose* by wearing a head-mounted display and vest of sensors and other digitalia. Media critic and theorist Mark Hansen describes experiencing the piece, how

> a forest clearing centering around a great old oak tree appears. Everything in your visual field seems to be constructed of light: branches, trunks, leaves, shimmer with a strange luminescence, while in the distance there appears a river of dancing lights. Leaning your body forward, you move toward the boundary of the clearing and pass into another forest zone. You are now enfolded in a play of light and shadow, as leaves phase imperceptibly into darkened blotches and then phase back again, in what seems like a rhythmic perpetuity. Exhaling deeply causes you to sink down through the soil as you follow a stream of tiny lights illuminating the roots of the oak tree.
>
> Soon you sink into an underworld of glowing red rocks that form a deep, luminous cavern beneath the earth. Exhaling again, you sink still further, encountering scrolling walls of green alphanumeric characters that (you will later learn) reproduce the 20,000 some lines of code upon which the world you are in is built. Longing for the vivid images above, you take in a deep breath and hold it, waiting to ascend. After passing once again through the clearing, you enter another world of text, encountering quotations from philosophical and literary sources that seem to bear directly on your experience. "By changing space, by leaving the space of one's usual sensibilities," one passage informs you, "you enter into communications with a space that is psychically innovative . . . we do not change place, we enter our Nature."

The attention you have been lending to your breathing makes you feel angelic and fleshy: while you float dreamlike, unencumbered by the drag of gravity, your actions are syncopated with your breathing in a way that makes your bodily presence palpable, insistent. Meanwhile, you find yourself floating back down to the clearing, no longer driven to explore, but meditative, content simply to float wherever your bodily leaning and breathing will take you. (107–9)

From Hansen's description one can see that "navigation" through *Osmose* depends on breathing: inhaling and holding your breath "moves" you up in the piece's world; exhaling moves you down. (Davies is a scuba diver, and she drew on her diving experiences in shaping how someone moves through *Osmose*.) Oliver Grau, who writes about new media art, lists how participants in *Osmose* described their sense of being immersed in a "contemplative, meditative peace" and of feeling "gently cradled" (199). Grau writes that *Osmose*'s "physically intimate design of the human-machine interface gives rise to such immersive experiences that the artist speaks of reaffirming the participants' corporeality; Davies even expresses the hope that a spatio-temporal context is created 'in which to explore the self's subjective experience of "being-in-the-world"—as embodied consciousness in an enveloping space where boundaries between inner/outer, and mind/body dissolve'" (199). Grau ends by noting that "Prerequisite to the attainment of this goal is immersion experienced in solitude, a subjective experience *in* the image world" (199).

Both *Osmose* and *Saturday*, as their creators hopefully describe in their quoted words, draw participants into unusual sensuous engagements with their environments and so are set up to encourage participants to attend to their hearing or breathing (in these particular cases) as they probably would not amid the distractive normalities of daily activity. Such attention to a body's sensuous perception characterizes many art pieces that rely on digital processing, such as *Ephémère*, another piece by Davies, or Paul Sermon's *Telematic Dreaming* (see Grau 274–75), Thecla Schiphorst's *Bodymaps: Artifacts of Touch* (see Hansen 64–67), or Elizabeth Diller and Ricardo Scofidio's *Blur Building* (see Hansen 178–83). By experimenting with art that is not experienced by a person sitting still before a monitor, digital artists can ask us to attend to senses other than or in addition to sight, to experience those senses so as to "extend the domain of sensibility for the delight, the honor, and the benefit of human nature," as Wordsworth wrote several centuries ago (qtd. in Abrams 395).

The theorists of new media art discussed here—Hansen, Grau, Anna Munster—give extensive descriptions of *Osmose* as they write about digital art that grows out of the visual tradition of European art—even if the art they are describing is no longer primarily visual in its appeal. Each draws on—overtly or not—traditional eighteenth-century notions of aesthetics to discuss the art. It is that focus that gets them—and digital art (because digital art is a highly academicized and intellectualized area right now, with theory being read by artists who in turn make art that moves the theorists)—into potentially awkward situations. These are the situations noted in the introduction, in which aesthetics and ethics break apart, situations we have been warned about at least since Walter Benjamin.

Part of the project for each of these writers is to legitimate digital art *as* art. As mentioned, the kinds of art discussed here—*Saturday*'s bone transducers and *Osmose*'s breath responders, for example—do not look like traditional two-dimensional or even three-dimensional visual art. Such art does not equate with an object like a stretched canvas or shaped stone, as a painting or a sculpture does; instead, as with *Saturday* or *Osmose*, the art is what one experiences while wearing mediating objects like gloves or vests. This art is highly technologized, requiring considerable time (and, often, space) for installation and testing before it can be shown—and such art certainly cannot just be hung on a wall or placed on a pedestal and left to the oversight of long-standing museum guards.

Some in arts institutions do resist this work: there are mainstream arts magazines whose writers do not discuss this art (for example, see *Art in America*); "museums have only begun to open their doors hesitantly to the art of the digital present" (Grau 10); and there are schools that refuse to teach its production. Such art cannot be sold as singular objects.

But this art is, of course, taught and displayed, often in new or expanded institutions; as Oliver Grau wrote in *Virtual Art* (2003), there are "new media schools in Cologne, Frankfurt, and Leipzig and the Zentrum für Kunst und Medientechnologie in Karlsruhe, Germany, is a heartland of media art, together with Japan and its new institutes, such as the InterCommunication Center in Tokyo and the International Academy of Media Arts and Sciences near Gifu. More recently, other countries, such as Korea, Australia, China, Taiwan, Brazil, and especially the Scandinavian countries, have founded new institutions of media art" (10). These institutions (as their technically oriented names suggest) are all fairly recent, however, and it is—in part—the work of writers such as those discussed here to publicize this work, create (understanding) audiences for it, and show that it fits or ought to fit within existing arts institutions with, if necessary, only slight modification to institutional practices.

And, of course, to show that something new is not really so new, one shows how it fits into tradition—which could be one reason the writers included here discuss such digital artworks in parallel with traditional aesthetic theory. This, of course, requires reshifting in the logics of the traditional—and so leads to the problems mentioned in the introduction. To flesh out these problems requires showing these writers' aesthetic turn.

For most people in the early twenty-first century, aesthetics cannot be understood except as historicized. As theory about evaluative judgments about art or other cultural productions, as theory about one's taste for Rembrandt or Thomas Kinkade, Mozart or Mariah Carey, aesthetics is, at best, considered descriptive of how particular people in particular temporal and geographical contexts feel pleasure in their engagements with certain kinds of objects. Among others in the twentieth century, Raymond Williams ("Taste is for Williams a name for the habits of the dominant class rendered as inherent qualities" [Shumway 104]) and Pierre Bourdieu (for whom "the 'aesthetic point of view' was the surest mark of class distinction" and "largely reducible to ideology, a form of political dominance" [Harkin 185]) have done much to establish current theories about "the business of affections and aversions, of how the world strikes the body on its sensory surfaces, of that which takes root in the gaze and the guts and all that arises from our most banal, biological insertion into the world" (Eagleton 13); they encourage us to an understanding that such theories can make no universal, eternal claims about bodies and senses.

In aesthetics' eighteenth-century origins, however, those who developed theories of aesthetics believed they *were* discussing universals and eternals. Aesthetics, as a named discipline, began (in most tellings) with Alexander Baumgarten's work in the mid-eighteenth century. Baumgarten took *aesthetics* from the Greek *aisthesis*, which (in the words of Martin Jay) "implied gratifying corporeal perception, the subjective sensual response to objects rather than objects themselves" (6). Questions of aesthetics were originally, then, questions about how we make judgments about our sensory relations to the worlds in which we move: Why do we judge something to be beautiful, sublime, disgusting? Kant argued that aesthetic judgments result when we understand how universal reason can resonate in our particular, individual sensuous takes on the world, through conceptual understanding. Under this telling (to quote Cassirer's interpretation of Kant), the Beautiful is a "resonance of the whole in the particular and singular" (318). Similarly M. H. Abrams describes how, with the rise of Romanticism in the late eighteenth century, "writers testified to a deeply significant experience in which an instant of consciousness, or else an ordinary object or event, suddenly blazes into revelation; the unsustainable moment seems to arrest what is passing, and is often described as an intersection of eternity with time" (385). In such tellings, aesthetic judgments are possible precisely because it was believed, first, that something universal or timeless inhered in what we judge to be beautiful or to be art and, second, that each person's bodily sensibilities gave the person visceral and so cognitive access to that universal or timeless thing.

At least three originary stories have been proposed for the appearance of aesthetics as a named field in the eighteenth century, as a named approach to thinking about certain kinds of experience. There are perspectives like M. H. Abrams's development of Carlyle's concept of "natural supernaturalism." Abrams describes how the eighteenth into nineteenth century:

Romantic era was one of technical, political, and social revolutions and counter-revolutions—of industrialization, urbanization, and increasingly massive industrial slums; of the first total war and postwar economic collapse; of progressive specialization in work, alterations in economic and political power, and consequent dislocations of the class structure; of competing ideologies and ever-imminent social chaos. To such a world of swift and drastic change, division, conflict, and disorder, the inherited pieties and integrative myths seemed no longer adequate to hold civilization together. (292–93)

The result, for Abrams, is that Romantic writers "undertook, whatever their religious creed or lack of creed, to save traditional concepts, schemes, and values which had been based on the relation of the Creator to his creature and creation, but to reformulate them within the prevailing two-term system of subject and object, ego and non-ego, the human mind or consciousness and its transaction with nature" (13). To use Martin Jay's phrasing of this genealogy, aesthetics—aesthetic feeling—became for the Romantics a way of "infusing the natural world with all the numinous meaning that had hitherto been reserved for transcendent spirit" (16).

In *Marxism and Literature*, Raymond Williams gives aesthetics its ground in (as one would expect) changing conditions of production and consumption; he argues that "it is clear, historically, that the definition of 'aesthetic' response is an affirmation . . . of certain human meanings and values which a dominant social system reduced and even tried to exclude. Its history is in large part a protest against the forcing of all experience into instrumentality ('utility') and of all things into commodities. This must be remembered even as we add, necessarily, that the form of the protest, within definite social and historical conditions, led almost inevitably to new kinds of privileged instrumentality and specialized commodity" (151).

Artwork, that is, is moved from church walls and windows onto easily transportable (and so marketable) frames, to be consumed in particular, subjective ways, as Williams describes in *Keywords:* "It is clear from this history that *aesthetic*, with its specialized references to art, to visual appearance, and to a category of what is 'fine' or 'beautiful,' is a key formation in a group of meanings which at once emphasized and isolated subjective sense-activity as the basis of art and beauty as distinct, for example, from *social* or *cultural* interpretations" (28).

Terry Eagleton understands the appearance of the notion of aesthetics as coinciding with a change in disciplinary practices: at a time when institutions of power were changing—as a merchant class took on decision-making facilities outside the realms of kingly disposition and governmental structures became civil instead of monarchical—the externally applied disciplinary constraints of monarchy could not hold. Aesthetics becomes a way for those constraints to become internalized and personal so that

> a vision could be projected of a universal order of free, equal, autonomous human subjects, obeying no laws but those which they gave to themselves. . . . What is at stake here is nothing less than the production of an entirely new kind of human subject—one which, like the work of art itself, discovers the law in the depths of its own free identity, rather than in some oppressive external power. . . . Power is now inscribed in the minutiae of subjective experience, and the fissure between abstract duty and pleasurable inclination is accordingly healed. . . . Kant retains the idea of a universal law, but now discovers this law at work in the very structure of our subjective capacities. (19–20)

Bringing oneself in line with the aesthetic tastes of the time was thus a way to bring oneself in line with "universal law" and order, with no applied external compulsion.

As mentioned earlier, I am not arguing for one genealogy over another, as though the genealogies were mutually exclusive. What matters here is the three qualities the writers I quote similarly note about the aesthetic theories of roughly two centuries ago: those theories directed attentions to intensified or heightened sensuous bodily perceptions—to aesthetic experiences, that is—as what connected particular bodies with something larger, ineffable, or at least inutile; as a result, in being so connected, one was to experience—viscerally—one's place in the ethical world, in the world of universal law governing how one was to live. In formulating such connection, the theorists made aesthetic experience "into an intense but solitary experience of the relationship between self and external nature" (Harkin 174), as the quotations from Williams and Eagleton suggest. Although neither the digital art nor the theories about it described at the beginning of this essay seek relationship between self and the ineffable, they draw on the other two aspects of the older theories: first, they can encourage the solitary, ahistorical, nonparticular, engaged experience at the core of eighteenth-century aesthetics—as with Davies' words about *Osmose*—and, second, current art and theories do attempt to tie aesthetic experience to the ethical, to one's relationships with others. These two aspects of earlier theories do not and cannot be made to fit back together when brought to bear on current understandings of sensing bodies in their worlds.

From the time of Kant, those who have studied aesthetics have tended to direct their attentions in three directions: toward the object conceived of as being worthy of aesthetic judgment, toward the judgment itself, or toward the aesthetic experience that links the sensation of the object with the judgment about it. As mentioned earlier, the digital art discussed here is problematic as object, and the case has to be made for these digital pieces to be worthy of judgment as art. And so it makes sense that the writers discussed here would focus on aesthetic experience—a heightening or intensifying of day-to-day perceptual experience—in any attempt to use aesthetic theory in legitimating digital work such as *Saturday* or *Osmose*. In so doing, they, like the eighteenth-century aesthetic theorists, hope to use aesthetics to make perception ethical.

Munster, in her 2006 book *Materializing New Media: Embodiment in Information Aesthetics*, uses the first four-fifths of her book to discuss what I would call the epistemological functions of new media art; in her last chapter she claims that "the aesthetics of technologically inflected, augmented and managed modes of perceptions is also about relations to others in the socius" (151), about, that is, our ethical relations with others. Here is Hansen's take (from his 2006 book *Bodies in Code: Interfaces with Digital Media*) on what digital arts can do: Because they engage our senses, but in unexpected or new ways, as *Saturday* or *Osmose* engage with our hearing or our breathing, such digital art pieces can

> broaden what we might call the sensory *commons*—the space that we human beings share by dint of our constitutive embodiment. This is because digital technologies:
>
> 1) Expand the scope of human bodily (motor) activity; and thereby
> 2) Markedly broaden the domain of the *prepersonal*, the organism-environment coupling operated by our nonconscious, deep embodiment; and thus
> 3) Create a rich, anonymous "medium" for our own enactive co-belonging or "being-with" one another; which thereby
> 4) Transforms the agency of collective existence . . . from a self-enclosed and primarily cognitive operation to an essentially open, only provisionally bounded, and fundamentally motor, participation. (20)

Similarly, Grau ends his 2003 book on virtual art by arguing that the "processes of digitization create new areas of perception, which will lead to noticeable transformations in everyday life" (347): "The roles that are offered, assigned, or forced on the users when interacting are an essential element in perception of the conditions of experience—experience both of the environment in a world transformed by media and of the self, which is constituted as never before from a continually expanding suite of options for actions within dynamically changing surroundings" (347).

Munster, Hansen, and Grau each make this eighteenth-century move: They use aesthetic experience as what enables us to move from perception to ethics. The writers ground ethics in epistemology through this way of teasing out aesthetic experience. They argue that what we know about the world through our senses (not necessarily at the level of the discursive) becomes the ground for opening up the potentials of how we live together, socially, ethically. Each starts with our individual perceptual engagement with the world and acknowledges that there are then social relations to follow—but how are we, in action, now, really to use intensified individual epistemological experience with digital art to then build or ground ethical relations with each other?

Hansen gives the fullest account of this move by mixing the phenomenological perspectives of Maurice Merleau-Ponty with Bernard Stiegler's considerations of technics. From Merleau-Ponty he takes the distinction of "body image" and "body schema." For Hansen, "body image characterizes and is generated from a primarily visual apprehension of the body" (39), the sense of body we have from seeing ourselves and others in mirrors or represented in photographs or film; as do other writers (see, for example, Shusterman's discussion of "representational aesthetics"), Hansen argues that conceiving of our bodies only or primarily through sight extenuates our potential as sensing beings. For Hansen, however, to open this potentiality is not simply a matter of giving the other senses the same weight as sight. Instead, Hansen argues, we need to reconceive of—learn to reexperience—ourselves through a body schema, which gives "priority to the internal perspective of the organism" (39) and which is therefore necessarily already embodied, already active within an environment; this is therefore a body "always in excess over itself" because it—its senses, including of itself—is not separable from but is instead constituted within (and constitutive of) its environment, "coupled to" its environment. Drawing on Stiegler's conceptions of technics, Hansen argues that "because such coupling is increasingly accomplished through technical means" (39) the digital art he discusses in *Bodies in Code* can help bodies experience their environmental coupling and so move us toward the "essentially open" ethical relations he describes.

> During Queen Victoria's state visit to France in 1855, there was an outcry at court, where the sensitive noses of the ladies thought they detected her wearing perfume containing a little musk. (Vroon, Amerongen, and De Vries, ch. 1)

This is, then, finally, where I focus on problems with trying to understand new digital art under two-centuries-old conceptions of perception and aesthetics. I question how Hansen's formulations might work, how we might get from perception to ethics, from experiencing *Osmose* or *Saturday* and enhanced "organism-environment coupling" to a transformed "agency of collective existence."

At the originary time of the notion of aesthetic experience, the link between epistemology and ethics was precisely what aesthetic experience explained. If your conception of ethics meant learning to understand and shape private experience in tune with universal patterns, then an aesthetic experience—an intensification of a day-to-day sensuous perception—was what made that linking possible: it made perception available for reflection and so helped you understand that your feelings were a microcosmic reflection of that universal order and that through your feelings you could unite yourself still more with that order. It did not matter that aesthetic experience was an isolated, solitary experience, because the experience was understood, precisely, to be what enabled you to experience the larger within you.

But given that our understanding of ethics does not now involve our learning to live with universal patterns, that possible aesthetic link between epistemology and ethics is broken. In addition, at the time of the development of the originary notions of aesthetics, sense experience was considered both private and natural: your individual sensing could link you with Nature because your individual sensing resulted from your natural being. Not long after the origins I have described, however, Marx argued that "the forming of the five senses is a labour of the entire history of the world down to the present" (qtd. in Stewart 59); more recently, much research in anthropology (think, for example, of the writings of Peter Stoller, Constance Classen, or David Howes) works to demonstrate that senses develop culturally and that different cultures and, within cultures, different social classes (think of Pierre Bourdieu's work) have different sensory regimes—and that our sensuous perceptions of the world do not just happen "naturally" but come to their shape in our varying, complex, and socially embedded environments.

We understand now that, within such environments, our senses are trained through repetition. Sensuous training happens simply through growing up: we are raised into the sensory patterns and habits of our culture, and the the training therefore seems to have never happened because it is simply part of the day-to-day of growing up or raising a child. As David Shumway, in "Cultural Studies and the Questions of Pleasure and Value," writes, for example, "Taste, it turns out, is learned, but, like language, it is easily learned at a particular age and as part of one's environment" (104). Sensory training, however, can also happen through aware and intensely repetitively patterned training, the "repeated, sustained engagement" that Hawhee, for example, demonstrates was the "shared trait of athletic and rhetorical training" for the Greeks (146) or that philosopher Richard Shusterman describes as accompanying the more contemporary body training of the Feldenkrais Method or Alexander Technique (154–81). Or consider, for example, the recent narrative of how young race car driver Colin Braun was taught by his father, a professional race engineer, "not only how to read" the data from heavily wired racing cars "but also how to correlate the traces with what he felt in the driver's seat" (Lerner 120). The younger Braun's "training regimen" "began at age 6, when he started analyzing data logged by a unit his father installed on his kiddie car. He learned to commune with his vehicles during tens of thousands of laps on a test track on the family's property. And he has spent countless hours hunched over laptops, deconstructing multicolored graphs of racing data in an effort to 'see' what his car is doing. 'I look at squiggly lines and know what they mean,' he says. 'I don't remember learning it. It's something I've always understood'" (117). Think, too, of narratives about the intense repetitive physical work people who have had strokes must undergo to relearn bodily movements (see, for example, Kawahira et al. or Luft and Hanley).

The digital art that Hansen, Grau, and Munster consider is most often shaped to emphasize isolated, individual, private experience. These writers talk about a participant's sensuous perceptions of the art as though the perceptions result not from how the participant's repetitious and socially sensuous history shapes her to perceive but rather from a single technologized event experienced in isolation. This is to hold onto, and perhaps encourage, an eighteenth-century notion of bodies.

That such a notion of bodily experience cannot now lead to ethically enabling aesthetic experience becomes poignantly clear if we consider some writing about an already-existing—and quite widely used—digital environment that encourages our use of digital technologies to explore, socially and repeatedly, the potential fluidity of sensory formations. It is not art that matters here but rather the Nintendo Wii gaming device.

The Wii encourages both individual and social play and is readily available, all demonstrated by any search for photographs tagged "Wii" at the Flickr photo-sharing Web site.[3] The Wii controller is different from previous game systems' joysticks or mouse controllers, which ask players to sit still while moving only their wrists to affect what happens on a screen; instead, the company that makes the Wii, Nintendo, "reimagined the controller, introducing a three-axis accelerometer that transforms your hand motions into in-game action, so you really *play* the games. In *Wii Tennis*, for example, swing your hand just as you would a racket. In *Excite Truck*, hold both ends of the controller as if it were a steering wheel" ("The Console that Gets You in the Game"). Because of its availability, its possibility for social play, and its engagement of a broader range of senses than sight, the Wii brings to particular focus my concern about any turn to eighteenth-century, sensuously based notions of aesthetics to understand—and also to shape—our sensuous engagements.

Walter Benjamin's concern about the aestheticization of politics in "The Work of Art in the Age of Mechanical Reproduction" was a response, in part, to the work of the Italian futurist Marinetti, who found, and argued that others should also find, beauty in violent action, including, ultimately, war. Benjamin argued that Marinetti's encouragements toward particular kinds of violence diverted the proletariat's energies away from acquiring property and so concealed fascism's attempts at controlling property ownership: "If," Benjamin wrote, "the natural utilization of productive forces is impeded by the property system, the increase in technical devices, in speed, and in the sources of energy will press for an unnatural utilization, and this is found in war." Benjamin understood Marinetti to believe, then, that war would "supply the artistic gratification of a sense perception that has been changed by technology" and that, therefore, humans "can experience [their] own destruction as an aesthetic pleasure of the first order."

Given the hours some will play computer games, we probably could discuss how gaming can distract from more than just distribution of property, but I am not going to try to connect the Wii gaming device—or *Osmose* or *Saturday*—with war or self-destruction, no matter how you judge current political structures and events. Rather, I want to take up a more general argument from Benjamin. Benjamin understood that Marinetti was attempting to move not from aesthetics to ethics but rather the opposite: he was shifting politics from being about ethical relations to being about heightened sensuous experience. For Benjamin, when ethics, epistemology, and aesthetics are weakly linked or even unlinked, such that political action is judged aesthetically just as readily as aesthetic experiences can be understood as having ethical weight, then any sense experience is worth intensifying and exploring aesthetically, even violence—with no grounding to connect it to any ethical placement or ramifications.

That lack of grounding underlies, I believe, the ambivalence that appears in a discussion forum of the online magazine *The Escapist*, in response to an article titled "It's Only a Wii Bit of Violence." The article considers how violent computer games such as Resident Evil 4 or Manhunt 2, if they were to be ported from other gaming systems to the Wii, would ask players to use the Wii controller to mimic onscreen violent action. The article's writer asks, "Is this a lack of imagination, or a conscious decision to omit violent mimicry? More important, should graphically violent games with conventional control schemes be rated more leniently than games that are less graphically violent but offer a more tangible connection to the violence via the control method?" (MacInnes). The questions imply discomfort concerning what we might experience wielding the Wii controller as we would a knife or a bat to kill. One of the eight commenters in the magazine's forum responded to the article (with all the grammatical quickness that can characterize such online conversations) by describing using the Wii controllers in precisely that way: "court12b: I'm trying to imagine the scenario. You turn around, see some psycho coming at you with a knife, your pulse quickens, flight or fight instinct kicks in, you raise your baseball bat at the same time your heart rate skyrockets, you start bashing away with all your REAL energy. blood sprays from his head as he collapses to the floor, time decompresses back to normal as you catch your breath."

The two responses immediately following court12b's initial comment are ambivalent.

Russ Pitts: I'm of two minds on this. On the one hand, I agree that once we've crossed that barrier between pushing buttons to create an on-screen action and actually mimicking that action, with the stimuli described, we've gone a step toward blurring the line between games and life. But the other half of my brains thinks this is awesome, and exactly what we've been clamoring for since we started playing computer games. And it thinks that we will still be able to tell the difference between games and life.

Archon: OMG—When does this game come out? That sounds awesome. I know, I know, I'm missing the political ramifications. I can't help it. This is the immersion I've always wanted.

The comments demonstrate how, then, intensified sensuous engagements that happen to be violent can be desired, and desired precisely because they are viscerally intense and pleasurable—because they can be, for these commenters, *only* aesthetic experiences, separated out from other aspects of our lives. Or rather, perhaps, not quite "only," for the commenters are ambivalent: they show social discomfort with the imagined beating of others at the same time they seek its felt-as-an-individual-body pleasure.

Perhaps the commenters' ambivalence results from experiencing one's sensuous perceptions as natural, discrete, asocial, and unlearned yet also understanding, discursively or not, that sensation nonetheless shapes one's social actions. In other words, the ambivalence could result from holding on to the two aspects of eighteenth-century aesthetic theory I described Hansen, Grau, and Munster using, but without having the conceptual bridge—the belief in something larger or ineffable—that allowed one in the eighteenth century to make the two aspects fit together such that isolated sensuous experience was no obstacle to ethical connection. Given that I am not going to argue for a return of the ineffable, it seems—if we want to use aesthetic experience to help us link perception to ethics—we would need to learn to be bodies that somehow perceive not alone but socially. That is, we would need not only to believe that our sense experience is the result of being raised within a particular social regime but also to experience having such an unnatural, learned body.

If I am fair to Hansen, his arguments—the four steps quoted previously—do seem to be trying to explain how digital art might change our sense of our bodies, but I hope I have made the case that the art cannot do this if it does not allow for intense repetition of its experiences or if it is shaped to emphasize an isolated body. Hansen does discuss one piece of art that addresses bodies experiencing art together, but his discussion eventually leads us back to the same problem of how we get from the epistemological to the ethical.

In his book *Bodies in Code* (2006), the last artwork Hansen describes in his chapter focused on perceiving bodies is *Body Movies*, by Rafael Lozano-Hemmer, a piece originally exhibited in 2001 in Rotterdam and since then shown in Lisbon, Linz, Liverpool, and Duisburg. Lozano-Hemmer describes the piece on his Web site: "*Body Movies* transforms public space with 400 to 1,800 square metres of interactive projections. Thousands of photo portraits taken on the streets of the cities where the project is exhibited are shown using robotically controlled projectors. However, the portraits only appear inside the projected shadows of local passers-by, whose silhouettes measure between 2 to 25 metres high, depending on how far people were from the powerful light sources placed on the floor of the square. A custom-made computer vision tracking system triggers new portraits as old ones are revealed." That is, participants on the streets or squares where this piece is projected must move between a series of lights and a wall to make the shadows into which the photographs of their fellow city dwellers are then projected; the participants' actions therefore control when photographs are projected. With *Body Movies*, Hansen describes for the first time digital art that has been designed for multiple participants; he writes that the art is "deployed in the service of a broader aesthetic aim—that of creating the possibility for a form of communion rooted in a technically facilitated kinesthetic space. . . . To this end, *Body Movies* expressly solicits collective participation and, through it, the emergence of unpredictable behaviors. As Alex Adriaansens and Joke Brouwer describe it, *Body Movies* invited people on the square, up to 50 of them at a time, 'to embody different representational narratives,' thereby allowing them to create 'a collective experience that nonetheless allowed discrete individual participation'" (101–2).

About *Body Movies*, Hansen notes that "in the words of one Dutch participant, there is a possibility for a 'strange kind of communication with people you've never met,' one where 'you're all together but you're also separate'" (101–2); as a result, Hansen argues that "Creating the possibility for such communion—for a truly impersonal communication or, better (following Walter Benjamin), for the 'communicability' that underlies and facilitates communication—is the ultimate aim, and the ultimate accomplishment, of Lozano-Hemmer's relational aesthetic" (102). Given that this is the last piece Hansen describes in the particular section of his book discussed here, and given that he describes *Body Movies* as "truly inspiring" (102), I assume that *Body Movies* must be the closest for Hansen to what makes possible the transformation of "the agency of collective existence" that he believes ought to follow from the digital art he describes. One implication of his approval of this artwork is that the ethical move he proposes must be made with art purposefully designed to emphasize social activity by engaging multiple participants together at once—which would seem to throw the ethical efficacy of art like *Saturday* or *Osmose* into question, since such art, as noted earlier, isolates participants both physically and experientially in their time with the art.

But, in addition, an apparent assumption behind Hansen's approval is that transformed agency will happen automatically, in simple response to the experience of social art like *Body Movies*. It is as though, for Hansen, strangers who move together to make shadows on a wall will necessarily understand something new and different about their sensuous engagement with the world and others. Is that understanding automatic visceral learning or is it discursive, encouraging participants consciously to choose to move differently with others ever afterward? In either case, the questions raised previously about the necessity of repeated experiences for learning new sensuous engagements still apply; in the second case, what in *Body Movies* would encourage participants to understand discursively—and in the terms Hansen wants—that their experience of the piece has given them awareness of the Benjaminian "'communicability' that underlies and facilitates communication"? Particular responses to sensuous experience depend on how one has been raised up into a sensuous body through sense training but in this case would also seem to require training about art—and, in the case of the Dutch participant Hansen quotes, it is the presence of documentary videographers, asking participants specific questions about their experiences, that resulted in the responses Hansen quotes, not the experience of the artwork alone. Without outside encouragement or training that prepared one to question the experience, any discursive link between the aesthetic experience and its ethical consequences cannot be presumed.

"To see things in a new way that is really difficult, everything prevents one, habits, schools, daily life, reason, necessities of daily life, indolence, everything prevents one . . ." (Stein, 43). What can we do, perceptually, to live well together? If we believe that how we understand—even experientially—our and others' skin, smell, or physical closeness affects how we live together, then perception always impinges on the ethical. My focus has been on whether aesthetic experiences—intensified or heightened perceptual experiences—can also affect how we live together by changing the structures of our sensing, by changing, therefore, even our understanding and so movements and uses of our bodies. Although I have questioned the efficacy of nonrepetitive aesthetic experiences, I do not want to dismiss the possibility that such experiences—like the transducing gloves described earlier—can be openings to reflection on or discussion about how our senses are shaped and so can be openings to critical understandings of how our senses shape our relations with others. I believe, however, that such openings are most likely to occur (like the comments about *Body Movies*) with encouragement, with the sort of questioning that comes with practiced and overt instruction.

We know that writing is always an effort with unpredictable effects—and nonetheless we study, teach, and apply rhetorical approaches. If we believe that our senses can be heightened and so perhaps changed by experiences composed for that purpose—if, in other words, we believe that our senses are persuadable—then rhetorical considerations should apply here, as well.

Although my focus has been on art and gaming, any text we compose engages us aesthetically. Written texts may be shaped to dull bodily sensation, or to emphasize cognition over sensuality, but this is only one way among many that we teach bodies what they are or should be. As I mentioned in my opening, recent changes in the technologies of texts can make the aesthetic possibilities of texts more obvious and more available to our rhetorical ends, and so I hope that this essay has persuaded that how we engage each other sensuously through our texts, any text we make, is worth discussion in our research and teaching as we query how we might bind our bodily perceptions with our ethics.

## Notes

1. For photographs of the gloves and how they are used, go to "Electronic and installation" projects at Raaf's Web site, www.raaf.org.

2. Screenshots of the piece are online at http://www.immersence.com/osmose/index.php.

3. As of December 2008, there were 640 groups at Flickr that include Wii photographs (http://www.flickr.com/search/groups/?q=wii), with at least 45 focused exclusively on the Wii and some having more than 900 members and more than 2,600 linked photographs.

## Works Cited

Abrams, M. H. *Natural Supernaturalism: Tradition and Revolution in Romantic Literature*. New York: W. W. Norton, 1973.

Benjamin, Walter. "The Work of Art in the Age of Mechanical Reproduction." UCLA School of Theater, Film, and Television. February 2005. Accessed 5 May 2008, http://www.marxists.org/reference/subject/philosophy/works/ge/benjamin.htm.

Blair, Carole, Marsha S. Jeppeson, and Enrico Pucci, Jr. "Public Memorializing in Postmodernity: The Vietnam Veterans Memorial as Prototype." *Quarterly Journal of Speech* 77 (1991): 263–88.

Bourdieu, Pierre. *Distinction: A Social Critique of the Judgment of Taste*. Translated by Richard Nice. Cambridge, Mass.: Harvard University Press, 1984.

Cassirer, Ernst. *Kant's Life and Thought*. New Haven, Conn.: Yale University Press, 1981.

Classen, Constance. *Worlds of Sense: Exploring the Senses in History and across Cultures*. London: Routledge, 1993.

Classen, Constance, David Howes, and Anthony Synnott. *Aroma: The Cultural History of Smell*. New York: Routledge, 1994.

"The Console that Gets You in the Game." *PopSci*. Accessed 10 May 2008, http://www.popsci.com/popsci/flat/bown/2006/product_67.html.

Dickinson, Greg. "Joe's Rhetoric: Finding Authenticity at Starbucks." *Rhetoric Society Quarterly* 32.4 (2002): 5–27.

Eagleton, Terry. *The Ideology of the Aesthetic*. London: Blackwell, 1990.

Fleming, David. "The Space of Argumentation: Urban Design, Civic Discourse, and the Dream of the Good City." *Argumentation* 12 (1998): 147–66.

Grau, Oliver. *Virtual Art: From Illusion to Immersion*. Cambridge, Mass.: MIT Press, 2003.

Halloran, S. Michael. "Text and Experience in a Historical Pageant: Toward a Rhetoric of Spectacle." *Rhetoric Society Quarterly* 31 (2001): 5–17.

Hansen, Mark. *Bodies in Code: Interfaces with Digital Media*. New York: Routledge, 2006.

Harkin, Maureen. "Theorizing Popular Practice in Eighteenth-Century Aesthetics." In Matthews and McWhirter, 171–89.

Hawhee, Debra. *Bodily Arts: Rhetoric and Athletics in Ancient Greece*. Austin: University of Texas Press, 2004.

Howes, David, ed. *Empire of the Senses: The Sensual Culture Reader*. Oxford: Berg, 2004.

———. *Sensual Relations: Engaging the Senses in Culture and Social Theory*. Ann Arbor: University of Michigan Press, 2003.

"It's Only a Wii Bit of Violence: Forum." *Escapist*, 15 May 2007. Accessed 10 June 2007, http://www.escapistmagazine.com/forums/read/6.41858.

Jay, Martin. "Drifting into Dangerous Waters: The Separation of Aesthetic Experience from the Work of Art." In Matthews and McWhirter, 3–27.

Kawahira, Kazumi et al. "Addition of Intensive Repetition of Facilitation Exercise to Multidisciplinary Rehabilitation Promotes Motor Functional Recovery of the Hemiplegic Lower Limb." *Journal of Rehabilitation Medicine* 36.4 (2004): 159–64.

Lerner, Preston. "Ultimate Geek Racer Drives by Guts, Instinct—and Algorithms." *Wired Magazine*, 25 September 2007.

Lozano-Hemmer, Rafael. "Selected Projects and Video." Accessed 10 May 2008, http://www.lozano-hemmer.com/eproyecto.html.

Luft, Andreas R., and Daniel F. Hanley. "Stroke Recovery—Moving in an EXCITE-ing Direction." *JAMA* 296.17 (2006): 2141–43.

MacInnes, Fraser. "It's Only a Wii Bit of Violence." *Escapist*, 15 May 2007. Accessed 10 June 2007, http://www.escapistmagazine.com/articles/view/issues/issue_97/542-It-s-Only-a-Wii-Bit-of-Violence.2.

Matthews, Pamela R., and David McWhirter, eds. *Aesthetic Subjects*. Minneapolis: University of Minnesota Press, 2003.

Mountford, Roxanne. "On Gender and Rhetorical Space." *Rhetoric Society Quarterly* 31.1 (2001): 41–71.

Munster, Anna. *Materializing New Media: Embodiment in Information Aesthetics*. Lebanon, N.H.: University Press of New England, 2006.

"Osmose." 2007. Accessed 8 May 2007, http://www.immersence.com/osmose/index.php.

Raaf, Sabrina. "Saturday." Accessed 10 June 2007, http://www.raaf.org/projects.php?pcat=1&proj=6.

Shumway, David. "Cultural Studies and the Questions of Pleasure and Value." In *The Aesthetics of Cultural Studies*, edited by Michael Bérubé, 103–16. London: Blackwell, 2004.

Shusterman, Richard. *Performing Live: Aesthetic Alternatives for the Ends of Art*. Ithaca, N.Y.: Cornell University Press, 2000.

Stein, Gertrude. *Picasso*. New York: Dover, 1984.
Stewart, Susan. "Remembering the Senses." In Howes, 59–69.
Stoller, Paul. *Sensuous Scholarship*. Philadelphia: University of Pennsylvania Press, 1997.
Vroon, Piet, Anton Van Amerongen, and Hans De Vries. *Smell: The Secret Seducer*. Trans. Paul Vincent. New York: Farrar Straus & Giroux, 1997.
Williams, Raymond. *Keywords: A Vocabulary of Culture and Society*. New York: Oxford University Press, 1976.
———. *Marxism and Literature*. New York: Oxford University Press, 1977.

# The Rhetorics of Online Autism Advocacy

## A Case for Rhetorical Listening

Paul Heilker and Jason King

Whatever else it might be, autism is a profoundly rhetorical phenomenon. Regardless of its cause or causes, regardless of whether we consider it a lamentable medical condition or an amazing demonstration of the value of human sociocognitive diversity, regardless of the disciplinary lenses and professional orientations we might bring to bear on the issues involved, whatever sense we can make of autism—and whatever actions we might take based on that understanding—result primarily from rhetorical activity. Let us begin by recalling Aristotle's ancient distinction between the necessary and the contingent. The proper domain of rhetoric, Aristotle said, is not the realm of the necessarily true, certain, or stable, but rather the realm of the contingent, possible, and probable (1357a, par. 4). Autism is rhetorical, in part, because we do not yet know what causes it, and we may not know for quite some time. In addition, there is considerable argument about what autism could possibly be, how we should think about it, and how we should respond to it. Is it a disease? A disorder? A disability? A diversity issue? All these things, and more? How meaningful—and to whom—are the unstable and contingent distinctions between autism, "high-functioning autism," Asperger's syndrome, and the increasingly popular diagnosis of pervasive developmental disorder—not otherwise specified (PDD-NOS)? Since our discourses about autism are fundamentally, pervasively uncertain, autism and rhetoric are thus deeply wed.

Another way that autism is rhetorical is that—whatever it is—it is being constructed and reconstructed in the public sphere via strategic and purposeful language use. As Dilip Gaonkar has said, "Rhetoric is the discursive medium of deliberating and choosing, especially in the public sphere" (8). Because all we are presented with in the public sphere is competing narratives and arguments about autism, all of which are clamoring for our assent and none of which are remotely disinterested, autism and rhetoric again are deeply wed. We offer here an introduction to a set of such rhetorics of autism vying for our attention and allegiance: online rhetorics of advocacy from two opposing groups of invested participants.

As the fathers of sons on the autism spectrum, our interest in the rhetorics of autism is personal as well as professional. We can say from both personal and professional experience that examining the rhetoric of those on the spectrum is a considerable challenge. Consider the following: Although our definitions of rhetoric are legion, what most have in common is their focus on language use in the social realm. Most definitions of rhetoric focus on the role of communication in social interaction. Kenneth Burke, for instance, writes that the "basic function of rhetoric [is] the use of words by human agents to form attitudes or to induce actions in other human agents" (41). Similarly Marc Fumaroli says, "Rhetoric appears as the connective tissue peculiar to civil society and to its proper finalities" (253–54). And Gerard A. Hauser maintains simply that "rhetoric is communication that attempts to coordinate social action" (2).

Similarly, even though our definitions of autism are also legion, what they, too, have in common is a focus on language use in the social realm, a focus on communication in social interaction. Indeed, two of the three primary descriptors of autism spectrum behavior, two of the three fundamental ways that autism presents itself in the world, have to do with communication in the social realm. The National Institutes of Health defines autism as "a spectrum that encompasses a wide range of behavior" but whose "common features include impaired social interactions, impaired verbal and nonverbal communication, and restricted and repetitive patterns of behavior" (par. 3). Likewise, the Centers for Disease Control and Prevention says "Autism spectrum disorders (ASDs) are a group of developmental disabilities defined by significant impairments in social interaction and communication and the presence of unusual behaviors and interests" (par. 1).

Unless one has direct and frequent contact with a person on the autism spectrum, the abstractions above are not particularly meaningful. Let us try to be more specific, then, by offering excerpts from the "tip sheet" Paul provides to his son's teachers at the beginning of each school year, those tips which focus on the *rhetorical* aspects of autism.

> Eli has Asperger syndrome, a condition on the autism spectrum. Like most people with Asperger syndrome, Eli
>
> —has difficulty with the unwritten rules of social behavior and is sometimes unaware of even simple social conventions;

—shows marked deficiencies in social skills, having difficulty sustaining conversations and sometimes making socially inappropriate responses to others;
—has difficulty developing relationships with peers his own age, preferring the company of adults;
—desires interaction with others but has trouble knowing how to make it work;
—has a strong desire to please others and an acute sense of embarrassment when he has erred socially;
—has difficulty reading and using nonverbal cues, like eye-to-eye gaze, facial expressions, gestures, and body language;
—has difficulty determining socially appropriate body spacing;
—finds it hard to gauge the emotional state of others and equally difficult to express his own emotional state;
—gets preoccupied with (and may monologue about) particular and idiosyncratic subjects of interest;
—has an advanced vocabulary, but is very literal, has difficulty processing figurative language, and is overly involved in repetitive language use, reciting stock phrases or lines drawn from previously heard material;
—tends not to "get" jokes, despite a strong interest in humor and jokes, especially puns.

Similarly, in a blogpost entitled "Note to Daycare Teacher upon Her Asking the Following Question," Jason's wife, Rachel, provides some tips.

John David has autism. So what does that mean?
  Sometimes it means a lot, and other times it means very little. First of all, it means that he has been slow to develop adequate expressive language. I think he has a lot to say, but he just doesn't know how to say it. So, instead, he screams and squeals. He also kicks sometimes, or tries to hit. Or, he says the only things he knows how to say—things he's heard on TV or movies, or that we've said to him. He can talk—he just doesn't always know what or how to say what he wants to. And that's frustrating to him, I think, and obviously to those around him. But please be patient with him and don't write him off. He will say what he needs to in some way or other. And if you're "listening" well, you'll know what that is. (King pars. 1–2)

In sum, working within these kinds of dynamics, that is, studying the communication and social interactions of a group defined specifically because of its difficulties in communication and social interaction, is a fascinating—though sometimes vexing—problem.

## Rhetorical Conflict between Autism Communities and Autistic Communities

The central dynamic that drives online discourse about autism is the conflict between *autism* communities and *autistic* communities and their contestation over who has the ethical right to speak for people on the autism spectrum. Since the mid-1990s, those who care for people on the autism spectrum have been using the Internet as a way to form support groups, share advice, celebrate victories, and commiserate about difficulties, among other things. Generally speaking, these *autism* communities consist of neurotypical parents and professionals who are working with children on the spectrum. The members tend to see autism as a disease for which they seek causes and cures (or at least preventive measures). Some of the more visible and powerful of these communities include Cure Autism Now! (which recently merged with Autism Speaks) and the Autism Society of America, which calls itself "the Voice of Autism." On the other hand, we have more recently seen the creation of online *autistic* communities, which, generally speaking, consist of adults and adolescents on the spectrum who reject the disease-and-cure model, advocating instead that autism is an inherent and inextricable part of a person's selfhood and that what is needed is not a cure so much as greater sociocultural acceptance and tolerance of this natural difference. In addition to Aspies for Freedom, for instance, we can look to other autistic communities such as Autistics.org, which calls itself "the REAL Voice of Autism."

Autism and autistic advocacy communities often exhibit very different goals and employ vastly different rhetorics. While "cure-oriented" *autism* advocacies deploy rhetoric in search of sympathy and awareness that will yield financial support for medical research and therapeutic interventions, *autistic* advocacy is "acceptance-oriented," seeking instead empathy and open-mindedness about autism as a different, yet acceptable, way of being. To this end, *autistic* communities work to dissolve what they see as negative stereotypes and stigma that mainstream culture associates with autism. In stark contrast, metaphors of tragedy and violence are common currencies for *autism* communities. Consider the following three examples, which have recently been the object of much discussion in online autism and autistic advocacy communities.

First, we can point to the recent fund-raising film *Autism Everyday*, which was produced by Autism Speaks for a fund-raising event but is now available (and highly popular) on YouTube. Director Lauren Thierry characterizes the film as an effort to "show the world what the vast reality [of autism] truly is" and to reveal the "dark and uncomfortable truths about living with autism." Thierry argues that a realistic picture of autism is needed to counter the "erroneous perceptions" that people might glean from stereotypes about "autistic savants" or from recent stories about a phenomenal basketball performance by a young man on the spectrum (Liss pars. 8–9).[1] Thus the film is developed through interviews that weave together the struggles and frustrations of five mothers of children on the spectrum. Thierry's orchestration is evident from the opening moments of *Autism Everyday*, as somber music and the shrieks, moans, and screams of children lead up to the opening credits. The remainder of the film follows suit as each of the five mothers laments the overwhelming demands and difficulties of raising a child on the spectrum. The most controversial moment occurs midway through the film as Alison Tepper Singer—one of the mothers being interviewed as well as an executive vice president for Autism Speaks—elevates the tragedy-rhetoric even higher: "I remember that was a scary moment for me when I realized I had sat in the car for about 15 minutes and actually contemplated putting Jody in the car and driving off the George Washington Bridge. That would be preferable to having to put her in one of these schools. It's only because of Lauren, the fact that I have another child, that I did not do it" ("Autism Everyday [Original 13 Minute Version]").

Although some, such as Thierry, see Singer's admission as "gutsy and courageous," many "acceptance oriented" advocates have not only harshly criticized her for her thoughts of violence but have also accused her of unequally valuing her daughters. Moreover, many have suggested that her statements point sharply to an insidious trend of sanctioning violence against people on the spectrum though "propaganda" that "demonizes" or "dehumanizes." As Ari Ne'eman, founder of the Autism Self-Advocacy Network, puts it, "A causal relationship exists between the rhetoric that reinforces a diminished value for autistic life and personhood and the implementation of that idea in the form of murder of people on the autism spectrum" (par. 34).

Although *Autism Everyday* has received much support and much criticism online, its representation of autism or autism parenthood is not uncommon. In fact, images of tragedy, violence, and evil are frequently associated with autism to build identification among parents of children on the spectrum. One particularly notable example is Susan F. Rzucidlo's "Welcome to Beirut" statement, published on the BBB Autism Online Support Network with the subtitle "Beginner's Guide to Autism." In fact, "Welcome to Beirut" is a doubly poignant statement: it not only compares autism parenthood to a violent and terrifying journey but also is based on a revision of Emily Perl Kingsley's famous 1987 essay, "Welcome to Holland," written about having a child with Down's syndrome. In Kingsley's essay, parenting a child with a disability is compared to a trip to Italy that is suddenly diverted to Holland. Despite initial disappointment, however, the Holland traveler soon learns to appreciate the windmills, the tulips, and "the Rembrandts" (par. 8). Rzucidlo's "Welcome to Beirut" appropriates the traveler metaphor for autism parenthood, but with a much different message: "One day someone comes up from behind you and throws a black bag over your head. They start kicking you in the stomach and trying to tear your heart out. You are terrified, kicking and screaming you struggle to get away but there are too many of them, they overpower you and stuff you into the trunk of a car.

Bruised and dazed, you don't know where you are. What's going to happen to you? Will you live through this? This is the day you get the diagnosis. 'YOUR CHILD HAS AUTISM.' There you are in Beirut, dropped in the middle of a war" (pars. 2–3). Although Rzucidlo writes that hope might be found in new medications, research, hard fighting, and new relationships, she cautions: "Don't get me wrong. This is war and it's awful. There are no discharges and when you are gone someone else will have to fight in your place" (par. 8).

Finally, as we have seen most recently, the vilification of autism even pervades medical communities. On December 1, 2007, the New York University Child Study Center, led by Dr. Harold Koplewicz, launched what it labeled a "Provocative New PSA Campaign" called Ransom Notes, which was a series of ads meant to alert Americans "to the silent public health epidemic of children's mental illnesses" ("Millions" par. 1). Each ad resembled a typed or handwritten ransom note that was signed by a neurological condition such as "Autism" or "Asperger's Syndrome." The autism ransom note contained the following ominous warning: "We have your son. We will make sure he will no longer be able to care for himself or interact socially as long as he lives. This is only the beginning . . . Autism" (par. 5). And the Asperger's syndrome note read, "We have your son. We are destroying his ability for social interaction and driving him into a life of complete isolation. It's up to you now . . . Asperger's Syndrome" (par. 7). John Osborne, president of the public relations firm BBDO, called the ad series "a wake-up call." He explains, "Left untreated, these illnesses can hold children hostage. That's why we've chosen to deliver our message in the form of a ransom note" ("Millions" par. 3).

In response the leaders of several autism advocacy communities launched an online petition that contained an open letter to the creators and supporters of the Ransom Notes campaign. Among other things, they argued that the campaign perpetuates stigma and "some of the worst stereotypes" by portraying people with disabilities as "kidnapped and possessed" ("An Open Letter" par. 1). Similar objections, many of them scathing, were leveled by other advocates. For example, one blogger on the "Whose Planet Is It Anyway?" blog wrote:

> One thing is clear: The repulsive attitudes expressed by Autism Speaks, BBDO, and Dr. Koplewicz do not represent the majority view. They do not speak for us. Rather, they speak for a tiny number of bitter, twisted New York elitists who feel angry and cheated about having children with disabilities in their otherwise glamorous families, and who are throwing a colossal temper tantrum and spewing their vicious prejudice all over the media. . . . They are a hate group, pure and simple. They share a hideous eugenic agenda. . . . They are enemies not only of the autistic minority population but of decent people everywhere. And it's past time for decent people to stand up in outrage and put a stop to their hate-mongering. ("Ransom Notes" par. 6–7)

Although many thought that the ads were "gripping and thought provoking" and productively defied "political correctness," the campaign was soon ended as a result of the public outcry of advocacy organizations and parents of affected children (Fritz 8).

As these examples illustrate, autism and autistic communities harbor very different foundational assumptions about what constitutes appropriate, productive, and ethical advocacy, which creates what Ne'eman calls a "zero-sum game" where "any positive attention about autism or autistics takes away from the hoped-for public pressure for a cure. By portraying people on the spectrum as tragic and all aspects of autism as horrible, more fundraising dollars can be raised.... That this ignores the reality of most autistic people—who are neither 'Rain Men' nor tragedies—is irrelevant to the ends sought by Autism Speaks and its fundraisers" (par. 49). As Ne'eman sees it, autism communities employ polemical discourse to make the strongest possible fund-raising arguments. Other autism communities, such as the BBB Online Autism Support Network, employ polemical rhetorics to foster identification among parents as they face the challenge of raising a child on the spectrum. And in the name of awareness, medical organizations like The NYU Child Study Center accost people in the hope of mobilizing them through the "visceral reaction" of fear for their children (Fritz 8). In each instance, however, autistic communities object to what they see as harmful caricatures that work to counter respect for and acceptance of people on the autism spectrum.

## Public Rhetorics of Autism

The binary here is obviously arch and deeply entrenched. But let us back up a moment to sketch the larger backdrop against which this current conflict is being played out. While the idea of autism as a rhetorical phenomenon may be new, the rhetorical effects of autism are not. Since many of the earliest diagnosed cases of autism involved individuals who did not speak, people who were therefore unable to actively represent themselves in traditional ways, people on the spectrum have historically been spoken for—and spoken about—by the neurologically typical, much as we are doing now. Their silence was a blank screen onto which we projected numerous fears, values, and misconceptions. We need think only of Leo Kanner's infamous "refrigerator mother" theory, first published in 1949, to appreciate the power and the longevity of the rhetorical constructions of autism and people on the spectrum in the public imagination. Such discussions of parental frigidity as a cause of autism moved fairly quickly from the limited audiences of professional journals to those of popular journalism: in 1960, for instance, in *Time* magazine, Kanner described the parents of autistic children as "just happening to defrost enough to produce a child" (qtd. in "The Child Is Father" par. 4). This theory that parenting influence is to blame for autism no longer enjoys the currency and support it once had, but it still has its contemporary manifestations and adherents. For example, consider the recent explanation of autism offered by Michael Savage on his nationally syndicated radio show: "I'll tell you what autism is. In 99 percent of the cases, it's a brat who hasn't been told to cut the act out. That's what autism is. What do you mean they scream and they're silent? They don't have a father around to tell them, 'Don't act like a moron. You'll get nowhere in life. Stop acting like a putz. Straighten up. Act like a man. Don't sit there crying and screaming, idiot'" (Aronow par. 1).

Similarly, in 1967, Bruno Bettelheim explicitly compared autism with being in a concentration camp, one such extended treatment ending with the following peroration, complete with an appeal to pity: "Here I wish to stress again the essential difference between the plight of these prisoners and the conditions that lead to autism and schizophrenia in children: namely that the child never had a chance to develop much of a personality" (68). The now ubiquitous idea that inside every person on the autism spectrum is a normal person trying to get out, that people on the spectrum are imprisoned within a shell of autism and struggling to break through it, would seem to be closely connected with Bettelheim's original argument. Its staying power is phenomenal.

Because what most people know of autism and people on the spectrum comes from arguments like these presented via mass media, some additional critique of those arguments seems in order. If, as Fumaroli suggests, rhetoric is the connective tissue linking people on the autism spectrum to the larger society, some examination of that rhetoric and its effects is clearly needed. On 21 March 2007, for instance, *Larry King Live* on CNN offered "The Mystery of Autism," a one-hour presentation we find iconic of the *autism* community's public rhetoric on autism and people on the spectrum. Epideictic rhetoric was on strong display, as celebrities such as Bill Cosby, Toni Braxton, and Gary Coles offered panegyrics for Suzanne Wright and her organization, Autism Speaks. Wright began her remarks with powerful ethical and pathetic appeals (complete with photographs) by noting that she began advocating for autism research when her grandson was diagnosed with the condition. Her grandson "went into the darkness of autism right before our eyes," she said. Wright invoked an urgent, agonistic rhetoric by voicing her strong support for the federal Combating Autism Act of 2006, and her belligerent stance was clearly marked not just toward autism, but also toward those who might not share her orientation on the issues involved. She then grew hortatory—while at the same time demonstrating a remarkable bit of audience analysis—in her call to action, contending that "we need to get the grandparents of this country galvanized" to help with autism advocacy because parents are too busy with the day-to-day difficulties, the frustrations, anger, and sadness of dealing with a child on the spectrum. Furthermore, the argument that autism is a childhood disease of epidemic proportions was foregrounded, with Dr. Ricki Robinson (also playing heavily on pathos) asserting that "more children will be diagnosed with autism this year than with AIDS, cancer, and diabetes combined."

In addition to looking at the means and the goals of individual utterances, we should also step back to look at the larger, perhaps less obvious, features of this discourse if we want to get a more fully useful sense of the public rhetoric on autism and people on the spectrum. Again, this episode of *Larry King Live* is iconic. On the one hand, we have Suzanne Wright, wife of Bob Wright, former chairman and CEO of NBC Universal, who is thus able to secure the help of three celebrities, an enormously popular journalist, and an entire cable news network to voice her particular message about combating this childhood epidemic. On the other hand, not a single person on the autism spectrum spoke during the entire broadcast, and autism in adults was never mentioned. The differential in rhetorical access, agency, and power here is quite remarkable. Whereas Autism Speaks got an hour of high-profile national media coverage (not to mention the inevitable reruns), people on the spectrum did not get to speak (and do not get to speak in mainstream media).

## Online Rhetorics of Autism

The place where people on the spectrum do get to speak about autism is the Internet, which is in many ways an ideal medium for them. First, the asynchronous nature of Internet textuality is a tremendous boon for people on the spectrum because they typically have difficulties monitoring and processing the enormous range and number of rhetorical and contextual cues that so powerfully determine meaning in face-to-face conversations. Whereas neurotypical people rapidly, effectively, and usually unconsciously monitor and process how an utterance's meaning shifts according to rhetorical factors such as the exigence, occasion, setting, and purpose of the utterance, each of which can be further complicated by the evolving motivations and roles of the various actors in the scene as it plays out in real time, people on the spectrum often struggle in their efforts to do so. In addition to trying to deal with this unfolding and dialectic complex of unspoken and nonverbal information, people on the spectrum also have difficulties accounting for the effects of a speaker's physical performance, that is, difficulties in processing how the meaning of an utterance changes according to a speaker's facial expression, tone of voice and inflection, body language, physical spacing from the audience, and gestures. For instance, a speaker's use of sarcasm, irony, and even mock-anger and mock-sadness may be obvious to the neurotypical members of his or her audience, but these ploys may be baffling or completely misinterpreted by people on the spectrum. Combine all of these processing difficulties with the social imperative to respond quickly in a conversation (or be thought to be rude or incompetent, or both), and the desire to withdraw from synchronous, face-to-face forms of communication becomes quite understandable. Many forms of Internet communication relieve people on the spectrum of the imperatives to track and respond quickly to the overwhelming, unwritten nuances of conversations that we find in face-to-face situations. People on the spectrum can use the Internet to read, interpret, reread, and reinterpret the statements of others over long periods of time, and then to craft and revise their responses and statements slowly and carefully, both of which are especially useful when engaging in charged encounters and argument.

Second, the Internet allows people on the spectrum to find one another, to collapse the time and space that would otherwise fragment them and perhaps make any larger forms of community impossible. Internet-based communities and movements such as Aspies for Freedom, Neurodiversity.com, and Aspergian Pride—as well as the ever-increasing, collective mass of the blog rolls listed by individual bloggers on the autism spectrum—constitute sympathetic audiences of like-minded peers, which work to reduce a writer's anxiety and thus increase the likelihood of his or her contributing to the conversation. Simply put, the Internet encourages rhetorical participation by people on the spectrum who are often functionally shut out of real-world, real-time conversations; it allows them to present their perspectives effectively when they may not have the communication skills to do so offline.

Online discourse has thus empowered people on the autism spectrum to become organized politically. Let us take the group Aspies for Freedom (AFF) and their online rhetoric as a case in point. The AFF was founded in 2004 by Amy Roberts and Gareth Nelson, and it is remarkable in its aggressive and agonistic efforts to claim authority over the discourses about autism. The name of the organization itself constructs its members as "freedom fighters," and its warlike rhetoric has been on display ever since its founding. For example, the front page of the AFF Web site offers the following manifesto: "We know that autism is not a disease, and we oppose any attempts to 'cure' someone of an autism spectrum condition, or any attempts to make them 'normal' against their will. We are part of building the autism culture. We aim to strengthen autism rights, oppose all forms of discrimination against aspies and auties, and work to bring the community together both online and offline" (par. 3). Early on, the AFF coined a new term to announce its agenda, "the autism rights movement," and in the time since, other groups have followed suit and coined their own covalent terms, such as "the autistic self-advocacy movement" and "the autistic liberation movement."

Similarly, the AFF quickly developed its own visual language to represent its perspective and counter the visual rhetoric of other groups. Autism communities were the first to seek visual symbols to represent their common understandings and collective efforts, and like many contemporary advocacy movements, they coalesced around the image of a ribbon, whether a physical one a person could pin to his or her lapel or a digital one for posting on their Web pages. Whereas a ribbon promoting breast cancer awareness is bright pink, for example, the autism community ribbon is imprinted with a jigsaw-puzzle design, and the interlocking pieces are tinted dark blue, light blue, red, and yellow. The puzzle motif foregrounds the mysterious nature of the condition; the varied colors for the pieces, those colors' dispersion across the visual field, and each piece's unique shape suggest the wide range of ways autism presents itself; and the interlocking design works to express the community members' desire for unity and support. The AFF, however, objected to the symbol's negative connotations—that people on the spectrum were enigmas, that they needed to fit in, that they had pieces missing, and so on. These connotations are even more pronounced in Autism Speaks' visual symbol: a single such blue jigsaw puzzle piece standing upright against a white background, an arrangement that clearly anthropomorphizes the piece's main body and five lobes into a representation of a human head, torso, arms, and legs. The AFF and other autistic communities have instead gravitated toward the use of a rainbow-hued infinity symbol (the horizontally oriented, elongated figure eight). This design extends the visual range from four distinct colors to the seven colors of the visual spectrum (and all the various shadings between the seven primary hues) to better represent the diversity of people on the spectrum, and it uses the graceful, perpetual curves of the closed infinity symbol to counter the negative connotations of the puzzle motif, stressing instead ideas of balance, fluidity, unity, and beauty.

The AFF's efforts to control the discourse about autism go far beyond reconstruing its visual rhetoric, however. For instance, within a year of its founding, the AFF had established June 18 as the international Autistic Pride Day, with numerous public events scheduled each year to promote its educational and political agendas. One such effort is the AFF's global attempt to have people on the spectrum recognized as an official minority group by the United Nations. On 18 November 2004, Amy Nelson used the Internet to publish an open letter to the UN, which began as follows: "We make this declaration to assert our existence, to be able to have a 'voice' on autism, rather than only that of experts and professionals in the field, to show how discrimination affects our lives, and that we want to direct a change from this type of bias against our natural differences, and the poor treatment that can ensue thereof" (par. 5). People on the autism spectrum, Nelson says, recognize themselves as a minority group on the basis of their linguistic, genetic, sociocultural, and behavioral similarities, a group that faces widespread discrimination—including well-documented cases of hate crimes and "mercy killings"—because of its members' innate differences and therefore needs increased legal protection. But the most courageous and remarkable claim made in this letter is that the autistic community is facing possible genocide, the eradication of its future generations. People on the spectrum,

Nelson writes, "are facing an imminent threat of possible cure, in whatever fashion that may transpire. Prenatal testing for autism could mean a form of eugenics, the total prevention of autism through genetic counseling before conception" (par. 12). Thus, much as Bettelheim did forty years earlier, Nelson suggests that the Holocaust itself is a proper analog for autism in society and the difficulties faced by those on the spectrum.

Given the overwhelming power differential that autistic communities face when dealing with autism communities, for instance, the vastly superior rhetorical skills and access that Autism Speaks continues to bring to bear, it is no wonder that the AFF and similar groups have adopted such a martial rhetoric of their own. Such warlike self-representation has surely served them well to galvanize and energize their constituencies. But what are the longer-term effects of their employing such a martial rhetoric and warlike stance? At what point does this self-empowering representation of people on the spectrum as warriors become unhelpful, even harmful, to those who invoke it?

We worry that this deeply agonistic rhetoric may do nothing more, ultimately, than make both sides dig in even deeper, make both sides ever more vicious in their attacks, ever less able to hear anything from the other side. For example, a particularly nasty flame war has erupted of late between members of the AFF and a blogger known as John Best. Members of the AFF have circulated a petition for Google to close down Best's E-Blogger site, entitled "Hating Autism," claiming that it "is in clear violations of the terms of service, as well as spreading bigotry and hatred against specific groups." One of Best's blog entries begins as follows:

Tuesday, July 03, 2007

Aspies for Freedom, Supernitwits

Aspies for Freedom is one of the spawns of Neuroinsanity. This is a group made up mostly of young people and a few older idiots like Phil Gluyas. The deranged philosophy that is seen here is the product of what Neurodiversity has done to corrupt youth into acting against their own best interests. These are potentially violent and dangerous people with mentally diminished capacities for reasonable thought.
. . . I have to wonder if the next serial killer we hear about will come from this group. (pars. 1–4)

We find ourselves torn here between being disgusted by Best's hate speech and vitriol, on the one hand, and, on the other, being deeply saddened by the AFF's efforts to *silence* him. There is something that seems especially wrong to us about *people on the spectrum*, about Aspies for *Freedom*, advocating that someone be silenced by force.

## A Case for Rhetorical Listening

What is needed are alternative tactics for fostering both understanding and productive rhetorics for autism advocacy. As Ne'eman asserts, despite their differences, parents and self-advocates share many goals and are all capable of making positive contributions. However, Ne'eman argues, those involved need to think strategically about negotiating the rift between the two communities, about how to "translate the autistic community's ideas" for "the wider world," and about how they might avoid rhetorical stances that lead only to "stalemate" (par. 51). As a first step, Ne'eman suggests that advocates refocus on articulating what autism is, rather than what it is not, as well as what autism advocacy aims to do, rather than whom it opposes (par. 52).

As one such effort to create an alternative discourse that fosters understanding rather than opposition, Eric Chen offers a rhetoric of reconciliation that might help move this increasingly static binary forward. On his I-Autistic Web site, he has begun what he calls an "Aspies for Forgiveness" campaign:

> I believe that the autistic community desperately needs an Aspies For Forgiveness campaign. . . . Autistics are upset that they have been rejected, bullied and trodden upon by non-autistics. Years of pain and endurance has manifested in many angry words like "don't you pity us" and "don't you dare make us normal." People with these angry words come together to make big angry organizations telling people how they despise ill treatment from nonautistics. But anger only makes more anger, unhappiness only creates more unhappiness. These emotions only bring people further from one another. No one has made peace through anger. It is not through reasoning or anger, but love and forgiveness, that calls upon peace. (pars. 1–4)

We can hope, then, that more useful and humane rhetorics of online advocacy may emerge organically from within the autism and autistic communities, but we submit that as rhetoricians, scholars of technology and disability studies, and writing teachers, we have an ethical and pedagogical obligation to help hasten such transformations, to intervene and educate those in both the autism and autistic communities so that their heartfelt but perhaps misguided and counterproductive discourses of advocacy issues may become more fully effective in achieving their ethical and political aims. Specifically we contend that an adaptation and deployment of Krista Ratcliffe's theory of rhetorical listening could do much to help alleviate the strife between these two polarized communities and help them begin to move beyond the current vexed stasis in their discourse. Rhetorical listening, we believe, offers concerned citizens, people on and off the autism spectrum, scholars, and educators the means by which we can all collectively begin moving toward the kind of forgiveness Chen is calling for, by which we can all respond to Ne'eman's call for a more strategic approach to bridging and healing the rift between the autism and the autistic communities.

Ratcliffe begins her book *Rhetorical Listening: Identification, Gender, Whiteness* by stating that her purpose is to answer Jacqueline Jones Royster's call for scholars "to construct 'codes of cross-cultural conduct,' that is, rhetorical tactics for fostering cross-cultural communication" (3). Ratcliffe calls her code of conduct, her tactic for fostering cross-cultural communication, rhetorical listening, which she defines generally as "a stance of openness that a person may choose to assume in relation to any person, text, or culture" (xiii). It is important to note that Ratcliffe's work focuses on issues of race and gender, but it is equally important to note that she foregrounds her project by suggesting how broadly one might conceive of the usefulness of her ideas: any person, text, or culture might be profitably approached using this stance.

Defined more specifically as "a code of cross-cultural contact," Ratcliffe says, "*rhetorical listening* signifies a stance of openness that a person may choose to assume in cross-cultural exchanges" (1). Finding ourselves in such exchanges, she notes, we are faced with a confounding question: "How may we listen for that which we do not intellectually, viscerally, or experientially know?" (29). Ratcliffe maintains, however, that when we engage in *rhetorical* listening, "*understanding* means listening to discourses not *for* intent but *with* intent—with the intent to understand not just the claims but the rhetorical negotiations of understanding as well." Thus rhetorical listeners can and should "invert the term *understanding* and define it as *standing under*; that is, consciously standing under the discourses that surround us and others while consciously acknowledging all our particular—and very fluid—standpoints" (28).

Such a radical invocation of openness is tremendously attractive, offering the possibility of liberation and growth on a variety of levels. Still, significant issues arise immediately. We must, for instance, understand and take into account that the "listening" we are proposing is metaphorical, because these encounters and exchanges between members of the autism and autistic communities would be happening via digitally mediated textuality. Just as there are powerful analogs between speaking and writing—as well as fundamental differences, which we lose sight of only at our peril—so, too, are there powerful analogs between listening and reading and fundamental differences as well. As we discuss rhetorical listening in this context, then, we are really talking about a particular form of rhetorical reading. Furthermore, as a code of conduct, rhetorical listening assumes at least three crucial things about potential participants. First, rhetorical listening is a trope for invention. If successfully invoked, it will produce perspectives and knowledge that are truly new, perspectives and knowledge that are, by definition, challenging if not alien to the participant's current, habitual, even entrenched ways of being in the world. As Ratcliffe puts it, rhetorical listening can help us "resist the coercive forces within dialectic/dialogue while remaining open to impossible answers" (8). Participants must thus be willing to embrace the possibility of the impossible. Second, rhetorical listening is a stance of openness. To be effectively deployed, it requires the participant to be truly open to these challenging and alien perspectives and knowledges, these impossible answers. This presupposes a great deal about the goodwill, motives, and intentions of the participants and assumes a significant desire to turn away from the current state of affairs, which are both highly charged and the source of whatever power the parties in the autism/autistic schism currently possess. Third, although we might more easily adopt such a stance of openness toward an individual person or text, the question of openness toward an entire culture is a very different matter of scale. Moreover, the debate about the existence of an autistic culture—that is, arguments for and against the existence of such a culture—lies at the very heart of the current conflict between autism and autistic communities. One cannot have a cross-cultural exchange if one does not believe the other culture even exists. In short, even in proposing rhetorical listening as a strategy to alleviate the strife between autism and autistic communities, we recognize that an array of substantive difficulties exist that will limit its usefulness. Some people may simply be incapable of adopting such a stance. Even so, we think these limitations can be turned around and employed as useful tests to identify potential first adopters, potential initial candidates for training and mentoring.

As Ratcliffe explains, rhetorical listening involves a fundamentally different way for participants to relate to—and relate *through*—discourse. It means transforming our "desire for mastery into a self-conscious desire for receptivity" (29). She cites Heidegger's claim about "the divided *logos* . . . we have inherited in the West, the *logos* that speaks but does not listen" (23). What we need to recover, she says, is a forgotten practice of "laying others' ideas in front of us in order to let these ideas lie before us. This laying-to-let-lie-before-us functions as a preservation of others' ideas . . . and, hence, as a site for listening" (23–24). This practice runs directly counter to our deeply habituated practice of listening in order to engage in dialogue or dialectic, that is, listening precisely in order to discover what we agree with or contest. Listening within a stance of openness, she says, "maps out an entirely different space in which to relate to discourse. . . . For when listening within an undivided *logos*, we do not read simply for what we can agree with or challenge. . . . Instead, we choose to listen also for the exiled excess and contemplate its relation to our culture and our selves" (24–25). It is through this fundamentally different relation to discourse, this entirely different space, this process of "letting discourses wash over, through, and around us and letting them lie there to inform our politics and ethics," as Ratcliffe suggests (28), that rhetorical listening can bring about change. As she writes, "Within this more inclusive *logos* lies potential for personal and social justice" (25).

Coming to inhabit this more inclusive logos is hardly an easy matter, of course, especially when the stakes are high and the parties have strong personal and emotional investment in the issues involved. A slow and careful approach toward the other is wise, and that alone is a difficult enough challenge. The distributed and asynchronous nature of online discourse should, however, be of considerable help in easing the difficulties of these initial cross-cultural encounters between members of the autism and autistic communities. Participants can approach the other as slowly as they like, backing off and returning to the encounter as often as they need to without feeling the pressure of the social imperatives of "politeness" and the like, which can bind people to uncomfortable face-to-face conversations in real time. In addition, utterances on stable Web sites can remain archivally available for extended periods of time and thus be approached again and again, as many times as is necessary to allow a participant to begin inhabiting a more inclusive logos.

Once potential first adopters (appropriately disposed first participants) have been identified, we would do well to teach them the value of two practices: what Peter Elbow calls "rendering" and what Ratcliffe calls "eavesdropping." Elbow's concept of rendering, of discourse that seeks to render experience rather than seeking to analyze or explain it, provides a useful way for us to begin recovering the forgotten practice of laying others' ideas before us to let them lie before us, which Ratcliffe notes is crucial in the creation of a more inclusive logos. Rendering runs directly counter to our deeply habituated practice of listening or reading precisely in order to analyze, explain, or argue, and thus most of us are ill prepared to do it. According to Elbow, although "discourse that renders is . . . one of the preeminent gifts of human kind," students are almost never asked to produce it. He notes that "virtually all of the disciplines ask students to use language to explain, not to render," and thus when "students leave the university unable to find words to render their experience, they are radically impoverished" (137). "To render experience," he says, "is to convey what I see when I look out the window, what it feels like to walk down the street or fall down," it is language that "conveys to others a sense of [the writer's] experience" (136). But rendering is not merely an autobiographical effusion, "not just an 'affective' matter—what something 'feels' like," according to Elbow. Rather, it is a prerequisite to an important kind of reflection and learning. As Elbow puts it, rendering "mirrors back to [writers] a sense of their own experience from a little distance, once it's out there on paper." To the extent that we "write about something only in the language of the textbook or the discipline," he says, we "distance or insulate [ourselves] from experiencing or internalizing the concepts [we] are allegedly learning." Thus, Elbow notes, "Discourse that renders often yields important new 'cognitive' insights such as helping us see an exception or a contradiction to some principle we thought we believed" (137). Even though Elbow's concept of rendering focuses squarely on the writer's experience of some phenomenon, we think it offers us a useful lever by which to help participants learn to render— rather than analyze or explain or argue about—the phenomenon itself. And it is this rendering of the phenomenon itself, this rendering of an opposing discourse, that seems essential in learning the skills of laying others' ideas before us to let them lie before us, skills crucial to the creation of a more inclusive logos.

In concert with teaching participants the value and the practices of rendering others' discourse, we would do well to help participants learn the value and practices of what Ratcliffe calls "eavesdropping" (104). Eavesdropping, she says, is "a rhetorical tactic of purposely positioning oneself on the edge of one's knowing so as to overhear and learn from others and, I would add, from oneself" (105). It is a courageous act, we think, and should be presented as such. Eavesdropping, Ratcliffe writes, involves "choosing to stand outside . . . in an uncomfortable spot . . . on the border of knowing and not knowing . . . granting others the inside position . . . listening to learn" (104–5, ellipses in original). It involves more than just a willingness to be open; it involves a willingness to be vulnerable, if not exposed. Moreover, as Ratcliffe notes, "eavesdropping requires an accompanying ethic of care" (105), that is, "being *careful* (full of care) not to overstep another's boundaries or interrupt the agency of another's discourse" (106).

Those people in the autism and autistic communities seeking to move forward out of the currently vexed and static discourse of their cross-community discourses could begin, then, by lurking, by simply visiting the others' online sites. But they would need to approach these sites with a fundamentally different attitude, stance, and purpose. With our guidance and mentoring, which could be provided asynchronously online through our own Web site devoted to this project, participants could be educated in the concepts and coached in the practices of eavesdropping and rendering. They could then attend to the others' discourse and attempt to render it in their own writing, honing laying-others'-ideas-before-us-to-let-them-lie-before-us practices, and they could send these renderings to us electronically so that we, as interested third parties, could both validate their renderings and offer suggestions for improving them. Participants' deeply entrenched predispositions to analyze, explain, and argue with another's discourse will surely take some time and assistance to grow beyond. But as rhetors and teachers, this kind of patient, repeated attending and responding and suggesting is precisely our skill set and our calling.

After a considerable amount of time has been spent rendering and eavesdropping on the others' Web sites and blogs, once participants have become comfortable with their laying-the-ideas-of-others-before-them-to-let-them-lie-before-them abilities, attuned to listening for the exiled excess, and articulate in rendering their digital discourse rather than seeking to engage with it argumentatively, then they might be taught to make the next critical move in rhetorical listening, which is to move beyond simply what is said to include why it is said, to move, in Ratcliffe's terms, from the claim to the underlying cultural logic. Rhetorical listening, she says, "invites listeners to acknowledge both claims and cultural logics. . . . By focusing on claims and cultural logics, listeners may still disagree with each other's claims, but they may better appreciate that the other person is not simply wrong but rather functioning from within a different logic" (33).

Once a participant has effectively rendered a significant corpus of the others' electronic discourse, he or she could be asked to revisit those renderings with a different purpose and lens. Seeking the underlying cultural logic of the others' discourse will mean using their renderings as a mirror, as Elbow puts it, a reflection that allows them to see how and where they have distanced or insulated themselves from experiencing how others experience the same phenomena of autism. Seeking the underlying cultural logic of the others' discourse in their renderings will mean actively and openly and purposefully seeking and embracing an exception or a contradiction to some principle participants thought they believed (Elbow 137) as they listen for the exiled excess and contemplate its relation to their culture and their selves (Ratcliffe 25). In helping participants move from rendering and eavesdropping to seeking the cultural logic of the others' digital discourse, the ostensible "move" is simple, though hardly easy: it means asking participants to move from what the other has said to why he or she has said it. Participants could revisit their previous renderings and offer speculative, essayistic, exploratory discussions about how and why another person might experience autism in these ways, about how and why another person might have come to think and feel about autism in these ways. Again, participants could send us these texts through the auspices of our Web site, and we could respond as interested third parties. But in responding to explorations of the others' cultural logics, our roles as rhetors and teachers will necessarily and explicitly begin moving into the roles of interpreters, intermediaries, emissaries, and mediators for both communities. Participants will trust us only to the extent that we maintain a firm adherence to a singular orientation and purpose, which is to help both the autism and autistic communities equally, to improve the overall quality and usefulness of all discourse about autism, and thus to help improve the lives of people on the spectrum, their caregivers, and the polis as a whole as a result.

## Conclusion

In conclusion we harbor no delusions about the difficulties we would face in taking up such a challenge for our pedagogical or rhetorical skills. To train members of the autism and autistic communities in the precepts and values of rhetorical listening would be to ask them to forego their current ways of being in the world, ways which have provided them with whatever agency, power, and success they currently possess, to ask them to be not only open to the impossible but vulnerable in the process, to ask them to begin a process that will never end. As Ratcliffe points out, "Rhetorical listening with the intent to understand, not master, discourses is not a quick fix nor a happy-ever-after solution; rather, it is an ongoing process" (33). But given that the polarized rancorous discourses between members of the autism and autistic communities are now affecting public discourse on autism as a whole, both online and offline, and given that autism will only occupy a greater place in public affairs, we think the time to begin this necessary process is now. Both the exigency and the means of addressing it are before us; we need only the will to begin.

We end here with a distinct sense of how the project we have traced is both a direct outgrowth of traditional rhetorical study and perhaps a fundamental shift away from it at the same time. On the one hand, we see our proposed project of training members of autism and autistic communities in the values and practices of rhetorical listening as a direct attempt to act upon the understanding of rhetoric we have learned from such scholars as Lloyd Bitzer. In "The Rhetorical Situation," for instance, Bitzer contends that "rhetoric is a mode of altering reality, not by the direct application of energy to objects, but by the creation of discourse which changes reality through the mediation of thought and action. The rhetor alters reality by bringing into existence a discourse of such character that the audience, in thought and action, is so engaged that it becomes a mediator of change" (4). By helping members of these sometimes viciously combative communities learn to eavesdrop and render and rhetorically listen to one another's discourse, we can help bring about the creation of a more ethical, humane, and effective discourse, which may change reality by mediating the thought and action of people both within and beyond these communities, altering how they perceive of, conceive of, and respond to autism and autistics. We can help bring into existence a discourse that may engage a wide range of persons to become more ethical, humane mediators of change themselves. But a project like the one we propose here also challenges us to grapple with questions about the appropriate sphere of rhetorical scholarship, both personally and collectively. What is the place of and for "socially responsible" scholarship in rhetorical studies? What is the place of and for *activist* scholarship in rhetorical studies? To what extent can and should scholars of rhetoric become mediators of public disputes between contending parties? What is at stake if we do? And perhaps more important, what is at stake if we do not?

### Note

1. High school student Jason McElwain, who is on the autism spectrum, received a flurry of media attention when he scored twenty points in four minutes during a high school basketball game on February 27, 2006. He was also honored by President George W. Bush. http://www.cbsnews.com/stories/2006/03/14/national/main1401115.shtml.

### Works Cited

"AFF: Aspies for Freedom." Accessed 15 June 2007, http://www.aspiesforfreedom.com.
"An Open Letter on the NYU Ransom Notes Campaign." PetitionsOnline. Accessed 22 July 2008, http://www.petitiononline.com/ransom/petition.html.

Aristotle. *Aristotle's Rhetoric: A Hypertextual Resource*. Translated by W. Rhys Roberts. Edited by Lee Honeycutt. Accessed 20 July 2008 http://www.public.iastate.edu/~honeyl/Rhetoric/rhet1-2.html#1357a.

Aronow, Zachary. "Savage on Autism." Media Matters for America. Accessed 6 December 2008, http://mediamatters.org/items/200807170005.

"Aspergian Pride." Accessed 15 June 2007, http://www.aspergianpride.com/.

"Autism Everyday (Original 13 Minute Version)." Autism Speaks. Accessed 22 July 2008, http://www.autismspeaks.org/sponsoredevents/autism_every_day.php.

"Autism Everyday–7 Minute Version." YouTube. Video. Posted 29 August 2007, accessed 22 July 2008 http://www.youtube.com/watch?v=FDMMwG7RrFQ.

"Autism Society of America." Accessed 15 June 2007, http://www.autism-society.org/site/PageServer.

"Autism Speaks, Home Page." Accessed 15 June 2007, http://www.autismspeaks.org/.

"Autistics.Org: The Real Voice of Autism." Accessed 15 June 2007, http://www.autistics.org/.

Best, John. "Aspies for Freedom, Supernitwits." Accessed 3 July 2007, http://hatingautism.blogspot.com/.

Bettelheim, Bruno. *The Empty Fortress: Infantile Autism and the Birth of the Self*. New York: Free Press, 1967.

Bitzer, Lloyd F. "The Rhetorical Situation." *Philosophy and Rhetoric* 1.1 (1968): 1–14.

Burke, Kenneth. *A Rhetoric of Motives*. Berkeley: University of California Press, 1969.

Centers for Disease Control. "Autism Information Center." Accessed 3 March 2007, http://www.cdc.gov/ncbddd/autism/.

Chen, Eric. "Forgiveness: Letting Go of Pain." Accessed 15 June 2007, http://iautistic.com/autistic-forgiveness.php.

"The Child Is Father." *Time*, 25 July 1960. Accessed 5 November 2008, http://www.time.com/time/magazine/article/0,9171,826528,00.html.

Elbow, Peter. "Reflections on Academic Discourse: How It Relates to Freshmen and Colleagues." *College English* 53.2 (1991): 135–55.

Fritz, Gregory K. "Mental Health in the Media: Both Sides of the Ad Controversy." *The Brown University Child and Adolescent Behavior Letter* 24.3 (2008): 8.

Fumaroli, Marc. "Rhetoric, Politics, and Society: From Italian Ciceronianism to French Classicism." In *Renaissance Eloquence: Studies in the Theory and Practice of Renaissance Rhetoric*, edited by James J. Murphy, 253–73. Berkeley: University of California Press, 1984.

Gaonkar, Dilip Parmeshwar. "Introduction: Contingency and Probability." In *A Companion to Rhetoric and Rhetorical Criticism*, edited by Walter Jost and Wendy Olmsted, 5–21. Oxford: Blackwell, 2004.

Hauser, Gerard A. *Introduction to Rhetorical Theory*. New York: Harper & Row, 1986.

Kanner, Leo. "Problems of Nosology and Psychodynamics of Early Infantile Autism." *American Journal of Orthopsychiatry* 19 (1949): 416–26.

King, Rachel. "Note to Daycare Teacher Upon Her Asking the Following Question." The Wonder Years. Posted 12 October 2007, accessed 22 July 2008, http://raking.squarespace.com/present/note-to-daycare-teacher-upon-her-asking-the-following-questi.html.

Kingsley, Emily Perl. "Welcome to Holland." *Our Kids*. Accessed 22 July 2008, http://www.our-kids.org/Archives/Holland.html.
Liss, Jennifer. "Autism: The Art of Compassionate Living." *Wiretap*, 11 July 2006. Accessed 22 July 2008, http://www.wiretapmag.org/stories/38631/.
"Millions of Children Held Hostage By Psychiatric Disorders." The NYU Child Study Center. Rubenstein Associates, Inc. Press release. Posted 1 December 2007, accessed 22 July 2008, http://www.mtnhomesd.org/mhhs/MHHSLibrary/Citing%20Sources.htm#Press%20Releases.
"The Mystery of Autism." *Larry King Live*. CNN. 21 March 2007.
National Institutes of Health. "Medical Encyclopedia: Autism." Accessed 25 February 2007, http://www.nlm.nih.gov/medlineplus/print/ency/article/001526.htm.
Ne'eman, Ari. "Dueling Narratives: Neurotypical and Autistic Perspectives about the Autism Spectrum." The Society For Critical Exchange. Accessed 22 July 2008, http://www.cwru.edu/affil/sce/Texts_2007/Ne'eman.html.
Nelson, Amy. "Declaration from the Autism Community." Accessed 15 June 2007, http://www.prweb.com/releases/2004/11/prweb179444.htm.
"Neurodiversity.com." Accessed 15 June 2007, http://www.neurodiversity.com.
"Ransom Notes and Autism Speaks: Partners in Crime." Whose Planet Is It Anyway? Posted 19 December 2007, accessed 16 July 2008, http://autisticbfh.blogspot.com/2007/12/ransom-notes-and-autism-speaks-partners.html.
Ratcliffe, Krista. *Rhetorical Listening: Identification, Gender, Whiteness*. Carbondale: Southern Illinois University Press, 2005.
Rzucidlo, Susan F. "Welcome to Beirut." Accessed 22 July 2008, http://www.bbbautism.com/beginners_beirut.htm.
Timelord. "Important Petition." Aspies for Freedom. Accessed 3 July 2007, http://www.aspiesforfreedom.com/showthread.php?tid=9597.

# Narrating the Future

## Scenarios and the Cult of Specification

John M. Carroll

It is a touchstone of contemporary culture that we invent the future. I was originally trained in the canon of early cognitive psychology and generative linguistics. But I have lived my career among people who do this literally—computer scientists, software engineers, and information technologists.

The activity of inventing the future is centrally about anticipating needs and interests of human beings. Not surprisingly, just what those future needs and interests will be is always unclear. When I was a graduate student, Alan Kay was inventing the Dynabook, the concept of a notebook form factor for personal computing. At the time, most people were still daunted by desktop terminals directly wired to gargantuan mainframes. Indeed, for years people scratched their heads about Alan Kay, but now it is clear that the only thing he really had wrong was the decade. He invented the future; it is our present.

In information technology, the Dynabook may stand out, but it is just a very sharp example of something that happens all the time. A more recent case is the MP3 player, most obviously, of course, the iPod. In a few short years, this device has been transformed from a slightly exotic item marking its owner as a cool, though most likely decadent, college student into a standard artifact of contemporary culture.

Technology is not only about increasing joy by fulfilling possibilities. It is also and often about perpetrating and then mitigating agonies. It is about managing problems. Technology development can be seen as iterating cycles of problem solving and problem spawning. At the time Alan Kay was dreaming about the Dynabook, many of his contemporaries were staring at blinking cursors, trying to guess what arcane command might coax the computer into making an understandable response. This was the so-called recall problem of early command line user interfaces. People are not especially good at recalling arbitrary command names, names of files, and complex labels. And recall difficulties are heightened when the recall target is downright cryptic, like the critical Unix command *grep* (for "search Globally for lines matching the Regular Expression, and Print them"; Carroll, *What's in a Name*).

Every problem in technology has many solutions. For example, the recall problem (among others) entrained a new sort of user interface that presented labeled document icons scattered about on a metaphorical "desktop." One accessed and manipulated the properties and the functions of these objects via menus. These innovations directly address the recall problem: People no longer had to recall file names or command names. They could rely on the graphical user interface to present these directly. They could rely on *recognition* memory instead of *recall* memory. People are extremely good at recognition.

Every solution eventually entrains its own boundary conditions. The graphical user interface seems like a great idea when one cannot recall a file name but can see the file plainly on the desktop. But after a while, when there are several hundred or several thousand files on the desktop, the idea is not quite so brilliant. In this way, each new solution entails a new set of problems, and further design and invention. I call this canonical pattern the "task-artifact cycle" (Carroll, *Making Use*).

In this essay I discuss scenarios—brief and evocative narrative descriptions—as a design representation. The bright side of my argument is that by describing and analyzing information technology designs through narratives of their use (that is, *before* those designs are ever implemented and deployed to users) future problems and possibilities can be anticipated and managed. The dark side of my argument is that traditional, specification-based design methods minimize the chance of anticipating problems or of achieving possibilities. From this, I conclude that information technology design should be construed and developed as a rhetorical practice and not merely as a systems and software engineering practice. We need to cultivate methods for narrating the future.

## Specifications

How can we get from imaginable possibilities and currently experienced problems to the future we want to invent? There is—always—an established approach, the establishment as it were. In the engineering disciplines, including computer science, software engineering, and information technology, the established approach is specification. A specification is a structured analysis of the parts and relationships that comprise a complex object.

A typical use of specification is functional specification, in which one enumerates the components and properties of a piece of functionality, for example, a command or a set of related commands in a software system. The functional specification defines a software component in terms of its parts and properties, its relationships to other bits of functionality, and information about how to operate it and how it is implemented. Clearly specifications are a useful sort of design representation, and they are critical documentation if one ever needs to repair, extend, or refactor an existing piece of software (as one nearly always does). Specification is used pervasively in the design of hardware, software, and even human activity systems, where it is called task analysis.

An example of a specification. It describes the Smalltalk inspector tool (Carroll and Rosson). The basic functions, the component parts and their properties, and the interactions with other Smalltalk capabilities are listed. A more detailed version of this specification would probably include a screen shot, because the inspector is a tool for a graphical user interface system.

---

### Functional Specification for a Smalltalk/V Inspector

The Inspector is a low-level debugging aid used to examine and edit objects.

Inspector components:
The *instance variable list* and the *instance variable contents*.

The *instance variable list* appears in a list pane positioned as the left pane of the tool. The first instance variable in the list is self, the object being inspected. The *instance variable contents* appear in a text pane on the right of the tool. This pane displays the contents of the variable currently selected in the list pane.

There is a special Inspector for Dictionary objects: the instance variable list contains Dictionary keys, rather than named or indexed instance variables. The special variable self is not included in the Dictionary Inspector variable list. There is also a special "method context" Inspector used by the system debugger (see below).

Inspector properties:
*Browsing the instance variable list:* Instance variables (including self) can be selected (clicked on) in the list pane; this causes the selected name to highlight and the contents of the variable to be displayed in the instance variable contents pane.

*Evaluation in the instance variable contents pane:* Message expressions typed in the instance variable contents pane can be evaluated.

*Evaluation under the scope of self:* Any of an inspected object's instance variables can be referenced in the message expressions evaluated in the instance variable contents pane.

*Save and update:* Expressions evaluated in the instance variable contents pane can be used to modify an object's instance variables; if an expression is "saved," the value of the selected instance variable will be set to the result of evaluating the expression.

Interface with other system tools:
A "method context" Inspector is incorporated into the system debugger. The object inspected in this case is the receiver of the currently selected method in the process walkback list. However, the instance variable list pane of this specialized Inspector displays on the receiver (self), the arguments to the method, and the temporary variables defined in the method.

Inspector functions:
*Inspection:* Inspection is initiated by sending the inspect message to an object (for example, via an expression typed and evaluated in a Workspace). Inspection can also be initiated by selecting an object and choosing Inspect from the Smalltalk menu. Within the Inspector tool, the menu for the instance variable list pane offers a single function Inspect, which opens a new Inspector on the currently selected instance variable.

*Text-editing:* The instance variable contents pane provides the standard text-editing menu, supporting the Restore, Copy, Cut, Paste, Show It, Do It, Save, and Next menu functions.

*Windowing:* The window menu for the Inspector contains seven standard window functions: Color, Label, Collapse, Cycle, Frame, Move and Close.

---

A key problem with specifications is that the representation is static. The object of design is defined and fixed when it is still just a plan on a piece of paper. Specification ensures properties of the plan—that it is comprehensive, closed; that known problems are addressed; that assumptions are enumerated. But ipso facto it leaves no room to maneuver, no room to explore and invent.

### Scenarios as a Focal Design Representation

During the past two decades, scenarios—narrative descriptions and envisionings of people interacting with technology—have become preeminent representations in software design. Of interest, the main historical root emanates from strategic planning, not design studies; perhaps this is the legacy of the new design methods. Also interesting is that the foundations and rationale for scenario-based design, although they are substantively cognitive, do not originate in cognitive science work on problem solving. They largely predate cognitive science and have developed independently of it. Nevertheless, it is interesting to reconstruct cognitive science foundations for scenario-based design. This could guide theory-based development of tools and techniques that might be overlooked on a purely practice-based understanding of scenario-based design. It could also help to produce a more generalized understanding of scenario-based design that might facilitate its application in other design areas.

Scenario-based approaches to strategic planning originated in work carried out in the late 1940s. The best-known example of this work is in Herman Kahn's

*Thinking about the Unthinkable.* Kahn developed the "accidental war" scenario: An incident of equipment malfunction or unauthorized behavior results in the launch of a single Soviet missile. The missile detonates in Western Europe, and this is immediately detected and disseminated throughout Western military and civilian installations. Although the incident is not interpretable as an attack, the level of anxiety throughout the Strategic Air Command results in one officer's misunderstanding or disobeying orders and firing the missiles under his command. The counterattack is also not interpretable as an all-out first strike; however, it is an escalated military response to the original mistake and could provoke the Soviets to launch their own ready missiles and bombers. In response, the United States might well order the rest of its missiles to be fired, as well as launching its bombers for protection from the Soviet assault and to position them for a subsequent response.

Not a nice story. Indeed, Kahn's planning scenario became a shared nightmare for much of the world for a quarter-century. It was also a very useful design aid. It is still shocking to know that in early 1961 no one had considered planning to communicate with the former Soviet Union during the five to forty minutes it would take for accidentally launched missiles to reach their targets. And yet doing so could possibly have saved the world had the scenario played out. *This possibility was identified by working through the accidental war scenario.* Kahn also developed extensions of this scenario in which the United States and the Soviet Union negotiated a limited nuclear war (this was later explored in the 1960s movies *Fail-Safe* and *Dr. Strangelove*) and in which they unilaterally established a world government (after surveying the first ten million casualties of Armageddon).

The accidental war scenario convinced Herman Kahn that a new regime was needed in strategic planning. He called the scenario a strange aid to thought— "strange" because, despite its shocking revelation, it was not conventionally employed. However, he argued that scenarios provide five distinct advantages to strategic planners.

1. Scenarios help analysts avoid the tempting assumption that circumstances will remain largely the same. They dramatically and persuasively emphasize the wide range of possibilities that must be considered.
2. Scenarios force analysts to address contextual details and temporal dynamics in problems, which they can avoid if they focus only on abstract descriptions.
3. By concretely illustrating a complex space of possibilities and imposing a simple linear rubric of time, scenarios help analysts deal with several aspects of a problem situation simultaneously. They facilitate the comprehension and integration of many interacting elements—psychological, social, political, and military.

4. Scenarios stimulate the imagination, helping analysts to consider contingencies they might otherwise overlook. They vividly illustrate principles or issues that might be overlooked if one considered only actual events.
5. Scenarios can be used to consider alternative outcomes to past and present crises.

The objective of strategic planning is to anticipate and assess future contingencies, resultant situations, and the courses of action they would present. The planner analyzes the problematic nature of current and future situations, situations that are at best only partially understood, and describes and evaluates possible responses. This is actually very similar to design, in which the objective is to envision future artifacts and to assess situations in which those artifacts would be employed in terms of personal and organizational consequences. Essentially strategic planners design courses of action to meet the requirements of future contingencies.

I have slightly extended Kahn's original analysis of scenarios in table 1, more explicitly contrasting the contributions scenarios can make to design to those more properly associated with specifications. Thus, as in the first row of the table, scenarios are deliberately sketchy, tentative, and malleable, whereas specifications are explicit, complete, and final. My point is not merely that one of these is better than the other, but that the two are vastly different and can be expected to make different, perhaps quite complementary, contributions to any design discussion or project.

Table 1

**Scenarios versus specifications as design representations**

| *Scenarios* | *Specifications* |
|---|---|
| Sketchy, tentative, malleable | Explicit, complete, final |
| Breadth-first thinking | Depth-first thinking |
| Temporal dynamics | Static structures |
| Raise questions; "what if?" evokes rationale/explanation | Derive/define answers; evokes entailments |
| Human activity/experience | Information/control flows |
| Exploits strengths of human problem solving | Controls flaws in human problem solving |
| Accessible to all stakeholders | Technical, arcane |

As shown in table 1, scenarios engage and encourage breadth-first thinking, whereas specifications evoke depth-first thinking. Traditional strategic planning and analysis methods were developed to explore enumerated possibilities, but they were not effective at identifying new possibilities. Scenarios emphasize temporal dynamics; specifications emphasize structural relationships that are basically static. Scenarios raise questions about why and how a narrative proceeds as it does; they provoke "what if?" thinking about alternative narratives; they cause people to seek and to provide narrative exegeses, explanations for their narratives. Specifications offer definitions and answers; they are designed to settle fundamental questions, not to leave them open or to open them more. The questions they provoke are quite focused questions of logical entailment, that is, the propositions that can be deduced from a given a set of asserted propositions (i.e., the specification). Scenarios depict human activity and experience; they encourage us to identify with the actors described in the narrative and to anthropomorphize the objects of the narrative. Specifications in contrast are about flows of information and control in systems and in user interactions with systems.

Scenarios deliberately seek to evoke empathy, imagery, and meaning making; people are *supposed* to feel something, to see and hear depicted events, to anthropomorphize objects, and to construct their own interpretations. Narratives are important cognitive archetypes of human thinking: we make lessons into folk legends, we dream, we share myths. Specifications are denotative; they definitely are not intended to evoke emotion or other subjective experiences. Indeed, this would be absolutely antithetical to the whole enterprise of specification. Scenarios explicitly seek to leverage the creative characteristics of human problem solving. Not only are people quite good at solving open-ended problems, but they also enjoy working on such problems.

Scenarios present open-ended design problems. In contrast, specifications try to address some of the known flaws in human cognition. For example, people have limited memories, so they often produce internally inconsistent solutions to complex problems. This can happen when they lose track of details, forget assumptions they have made or details of a subsolution, and then go on to address further aspects of a complex problem. Specifications address this by comprehensively codifying problem solutions: indeed, tools and languages for specification ingeniously hide complexity, for example, presenting a top-level view whose components can be successively expanded.

Finally, scenarios are accessible to all stakeholders in a design—users, managers, customers, clients, and even their relatives. A big part of the extraordinary impact of Kahn's accidental war scenario was that everyone could understand it and its significance almost immediately. Specifications, like many technical representations used by professionals, are inherently arcane to nonprofessionals. This is understandable and perhaps necessary, but it is also unfortunate. When all stakeholders cannot participate directly in design, they cannot share their perspectives and knowledge, and the design process itself is undermined (Carroll, *Making Use*).

My first professional position was as a scientist, and later a manager, at IBM. I worked for IBM from the mid-1970s to the mid-1990s. The first functional specification documents I saw were humongous encyclopedias of features and functions (and this was before IBM discovered fonts other than Courier). I noticed that after the culture of human-computer interaction became established in IBM in the early 1980s, interaction scenarios started to appear as appendices to these documents, illustrating how the features and functions specified worked together to produce what we would now call a user experience. By the end of the 1980s, the scenarios had moved up front: The encyclopedic functional specification had become appendix to the sketchy user-oriented scenario. The rhetorical was beginning to lead; the purely functional was beginning to play a supporting role.

### Example: A Virtual School

The next two sections illustrate how ostensibly mundane scenario-based practices can be salutary. In the first example, crafting a scenario helped us recognize that we were about to build the wrong system; this is probably the most common mistake in systems design. In the second example, crafting a scenario opened an inclusive communication space for the entire design team, which helped us to allocate roles, creativity, and power more equitably.

When I left industry to become a professor in 1994, I decided to actively explore "alternative" paradigms for system design. I focused much of this effort on projects in the public and civic sector, working directly with teachers and community leaders. I was interested in seeing how scenarios could change software design as I had come to understand it during eighteen years at IBM.

My first major project was carried out under the National Science Foundation's Networking Infrastructure for Education initiative, announced in 1993. This program had a variety of exciting and challenging goals:

> Exploit the Internet in public education;
> Support collaborative, project-based classroom education;
> Increase student achievement with respect to standards;
> Improve access of rural schools to new educational opportunities;
> Enhance gender equity in science and mathematics education; and
> Reduce equipment costs and handling in science education.

This was a highly successful initiative in many ways. The list of goals is a hodgepodge, a wish list of goals, many of which are just as urgent today as they were then.

I led a group at Virginia Tech in proposing a rural infrastructure to connect science classrooms. We argued that this could leverage teachers and equipment across distances, allowing classes to be offered in schools where they rarely could be offered, and enhancing the critical mass of science classes more generally.

Our proposal included a problem scenario describing the challenges that a young girl might experience trying to learn physics given the constraints of her rural school and home. We also developed an envisionment scenario, sketching how some of these challenges might be ameliorated if the girl could have better access to resources, including other students, through the Internet. A shortened version of this scenario is shown here.

### Solar System Experiments in a Virtual School

Marissa is a student in Ms. Browning's physical science class at Blacksburg Middle School. The class today was about the solar system. After school Marissa wonders how gravitational relations would work if large gas giants such as Jupiter were much closer to the sun. She accesses Ms. Browning's virtual lab through her Web browser. Randy and David are already there, and she poses her question to them. The three students use a solar system simulation, manipulating various parameters. At the end of the two hours, the students shut down the lab to go on to other work. But before leaving, they agree to log on that evening at seven o'clock to review their findings and write up their report.

---

This scenario was engaging to National Science Foundation reviewers, as well as to Virginia Tech administrators, who created a remarkable publicity collage depicting it (fig. 7.1) for its new interdisciplinary science center. (I still love this vision: Take an after-school snack and a laptop out to a pasture, somehow pick up a broadband wireless network, and begin exploring the foundations of physics. Perhaps the solar system will just drop in for a visit!)

As we began working with real teachers and students to plan the design and implementation of our virtual school, we experienced directly how scenarios can facilitate critical analysis of whiz-bang technology ideas. It is of course *possible* that a group of students would meet in a virtual environment and engage in revelatory self-initiated investigations of gravitational relationships in solar systems. It is also possible that they would get stuck, or side-tracked, or seriously confused, or just exchange social chitchat. Indeed, since the mid-1990s all of us have seen that the Internet can be a powerful educational medium and resource, but that for teenagers it is often more readily appropriated for social interactions. These are important uses of the Internet, but they are not about doing creative physics.

The teachers and students we worked with on this project actually recognized these issues from the scenario analysis, and they helped our team alter its plan. We became aware of this process through a series of participatory design workshops with teachers in which we analyzed videotaped classroom activities (depicted scenarios of science pedagogy) and deconstructed the causal relationships among the various things students and teachers were doing and the kinds of outcomes they wanted to achieve—and to avoid.

Fig. 7.1 Virginia Tech publicity collage based on the virtual school envisionment scenario

We never implemented the scenario depicted in sidebar 2 and figure 7.1. Instead, we built a highly advanced (for the time) collaborative environment that focused on supporting classroom-to-classroom collaboration, as well as classroom-to-community expert interactions, guided by teachers who helped keep students on track with respect to learning science. Our vision eventually evolved into one of leveraging resources to help make the whole greater than the sum of the parts on a regional basis in high school science education (Isenhour et al.; Carroll et al.). We were fortunate that our original envisionment scenario was not a specification, and we treated it as the starting point for a design discussion.

### Example: Community Informatics

A second example, a more recent design study, is the Civic Nexus project, also supported by the National Science Foundation. In this project we tried to understand challenges and design effective and sustainable interventions for technology learning and management in community nonprofit groups, such as local historical societies, food banks, water quality groups, after-school enrichment programs, arts groups, senior citizen groups, regional emergency planning groups, low-income housing groups (for example, Habitat for Humanity), and churches.

The nonprofit sector in American society is fascinating. Although the very name, "nonprofit," emphasizes a sort of economic agnosticism, nonprofit organizations make a huge contribution not only to the economy but also to the social fabric of society, providing social support and social capital on which the coherence of the society depends. Indeed, in the United States much of the social welfare apparatus is implemented through the nonprofit community. Almost 6 percent of all U.S. organizations are nonprofit, accounting for more than 1.6 million organizations and 9.3 percent of all paid employees. The Johns Hopkins Comparative Nonprofit Sector Project estimates that in the late 1990s, in the thirty-five countries worldwide participating in its study, this sector had aggregate expenditures of $1.3 trillion and employed, when factoring in churches, 39.5 million full-time-equivalent workers (Salamon, Sokolowki, and List).

One organization we worked with, a sustainable development group, had an interesting identity problem with its Web site. I was enthusiastic about working with the organization initially because I found the Web site design to be clean and evocative (fig. 7.2). It was easy to read, restrained with respect to blocks of text, and had a graphical environmental theme of green imagery. Indeed, I met my collaborators in this group by browsing Web sites of nonprofit organizations in Centre County, Pennsylvania. I was quite surprised when I finally got to meet these folks face-to-face and learned that they did not want to be perceived as "mere tree huggers."

I learned something from this organization about sustainable development. It is not just a matter of protecting trees, wildlife, and streams; it is about the whole environmental system. It is about balancing economic development with environmental integrity—hence, *sustainable* development. This is, of course, much more ambitious than just restraining economic development and waste in favor of preserving what is natural and clean.

The organization had hired a consultant to rework its Web site. He had misunderstood their identity and mission, much as I had. He had produced a Web site that conveyed this misunderstanding pretty effectively. Indeed, he liked the Web site he produced so much that he refused to change it, and was in effect holding the organization's Web content hostage. When I met them, the organization's members lacked not only the skills to take back control of the Web site design but also the data that their site displayed, for example, water quality data and maps.

Through the course of our collaboration, we emphasized that detailed domain knowledge is a critical asset in design. In this case, knowledge of sustainable development was critical to getting the "right" design. The consultant did not have this knowledge, and although he had solid Web development skills, his knowledge deficit led him to design an attractive but misleading Web site. We helped the staff revalue the importance of their own design knowledge by suggesting that they write scenarios describing the kinds of experiences they wanted their users to have and the way that their Web site information would evoke these experiences.

Doing this literally took just one hour (although fully appropriating a scenario-based practice took a few months). But in just the first week we had a major breakthrough: One member of the organization wrote a scenario in which a local official comes to understand how wastewater management is part of the

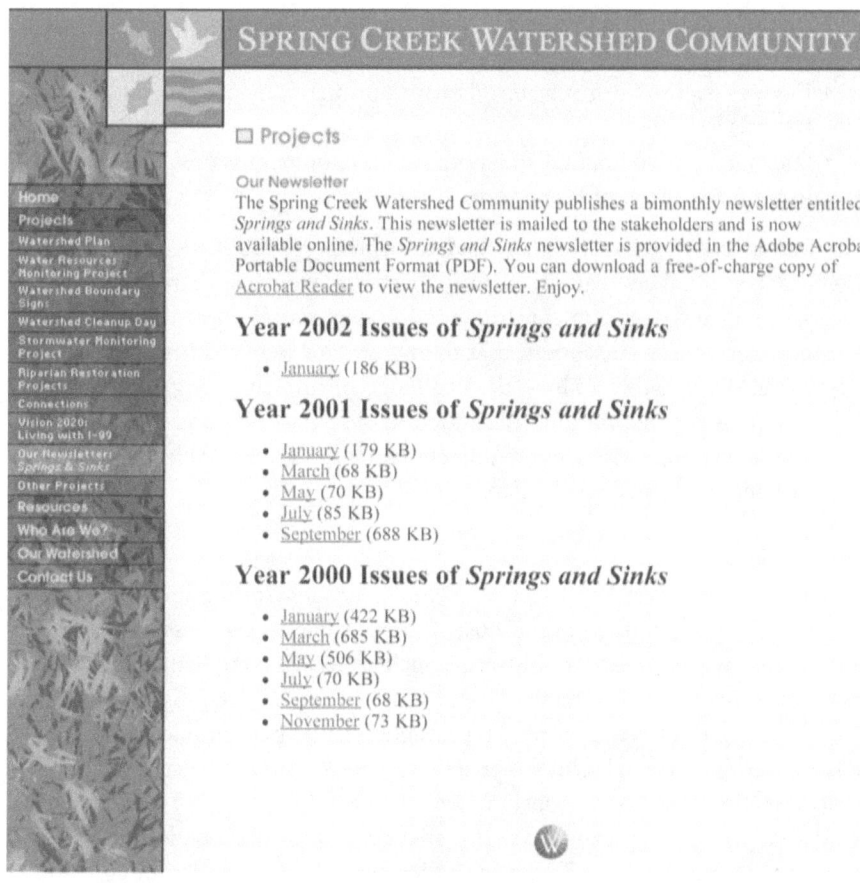

Fig. 7.2 Web site of Spring Creek Watershed Community in 2003

overall environmental system; following is a schematic representation. Prior to this scenario, the group had not focused on the special user group of local political leaders. So this really was an insight, both for Web-site design and for the group's identity in the community. Members of this sustainable development organization could not specify their Web site, but they were easily able to describe it through scenarios, and in fact to elaborate and discover design ideas through scenario envisionment. This allowed them to take control of their Web site through the mastery they already had of their domain and their Web site's content.

Elected Official Design Scenario Created by a Member of Sustainable Development Group

- Newly elected official has heard about SCWC, wonders what is it?
- Googles; browses site

- Pictures of local quality of life, mission statement, strategic goals and stakeholders
- Pursues topic of storm water: damages property, businesses, and local streams
- "Gets it": environmentally responsible development is cost-effective in both the short and long term

This reframing of what design is had a surprising effect with respect to the organization members' earlier attitude of powerlessness with respect to Web programming and development. It seemed that once they convinced themselves that they were the key designers, that their scenarios were the design, they were able to more effectively address the challenge of implementing and managing their design. A few months after the adoption of scenario-based design practices, some members supervised the implementation of a new Web site and then began to personally manage it (Farooq et al.).

## Conclusion

I have suggested that, by describing and analyzing information technology designs through narratives of their use, future problems and possibilities can be anticipated and managed. In this sense, such scenarios complement traditional, specification-based design methods, which are strong on detailing static properties of a design but weak on helping designers anticipate problems or achieve possibilities. The virtual school scenario (sidebar 2) and the elected official scenario just discussed are obviously more modest schemes than Herman Khan's accidental war conceptualization. But managing global crises is not the only thing people need to do. These examples show how everyday design breakthroughs are facilitated by scenario-based approaches. Indeed, one point I take from them is that, in the information age, inventing the future may become an everyday task.

Information systems design is in part a software engineering practice. And technical implementations should be specified, if only to create a detailed record of what was done. But the most important business of design always lies in the future, in envisioning and developing possibilities, and in anticipating and managing problems. Rhetoric can be dismissed as "just talk," but scenario-based approaches show how talk is constitutive of design. I believe that architects of technology from all disciplines must learn to narrate the future, and rhetoric obviously has a central role to play in this.

## Works Cited

Carroll, John M. *What's in a Name: An Essay in the Psychology of Reference*. New York: W. H. Freeman, 1985.

———. *Making Use: Scenario-Based Design of Human-Computer Interactions*. Cambridge, Mass.: MIT Press, 2000.

Carroll, John M., and Mary Beth Rosson. "Human-Computer Interaction Scenarios as a Design Representation." In *Proceedings of the 23rd Hawaii International Conference on Systems Science*, 555–61. Los Alamitos, Calif.: IEEE Computer Society Press, 1990.

Carroll, John M., et al. "Knowledge Management Support for Teachers." *Educational Technology Research and Development* 51.4 (2003): 42–64.

Farooq, Umer, et al. "Participatory Design as Apprenticeship: Sustainable Watershed Management as a Community Computing Application." In *Proceedings of 38th Hawaii International Conference on Systems Science*, 178. Los Alamitos, Calif.: IEEE Computer Society Press, 2005.

Isenhour, Philip L., et al. "The Virtual School: An Integrated Collaborative Environment for the Classroom." *Educational Technology and Society* 3.3 (2000): 74–86.

Kahn, Herman. *Thinking about the Unthinkable*. New York: Horizon Press, 1962.

Salamon, Lester M., S. Wojciech Sokolowki, and Regina List. *Global Civil Society: An Overview*. Baltimore: Johns Hopkins Center for Civil Society Studies, 2003.

# 3

Understanding Writing and Communication Practices

# Technology, Genre, and Gender

## The Case of Power Structure Research

Susan Wells

We know that gender affects both access to technology and the practices of users of technology, and we know that genres are associated with gendered practices of reading and writing. We know that new technologies foster new genres and that new genres have emerged with the development of digital technologies—blogs, wikis, and podcasts, to name the most familiar (Miller and Shepherd). We know relatively little about how this knot of association is structured: what are the theoretical relations among gender, technology, and genre? How do these relations change at moments of political or cultural crisis? This essay is a modest effort to see whether the concept of affordance might connect issues of gender, technology, and genre as they operated in the 1960s and 1970s. Then, the "power structure research report" emerged as a genre, mobilizing the affordances of photo-offset printing; the affordances of power structure research would themselves be appropriated by the new feminist movement.

"Affordance" is a term used and disputed in science and technology studies. Imported from psychology, it was invented by James Gibson to describe relationships between an environment and an animal: "the affordances of the environment are what it *offers* the animal, what it *provides* or *furnishes*, either for good or ill" (115). For theorists of technology, especially those interested in design, affordance pointed to the relationship between the design of a technology and the activities that it constrained or encouraged (Gaver; Pfaffenberger). The term is controversial: some theorists consider it too loose to be useful; others hold that it naturalizes technologies and their uses (Oliver; Hutchby, *Conversation*; Rappert; Hutchby, "Affordances"). But the term is also both flexible and widely used; scholars of multimodal literacy have imported it into rhetorical studies (Jewitt and Kress). Affordance is a mobilizing concept that orients us to action and interpretation as they play out in the materials of production. In the case of photo-offset printing and the power structure research report, the affordances of technology and genre serve as reflexive representations of each other to readers and writers.

Affordance might therefore be a link between gender and genre. The case of offset printing and the genres it supported in the 1960s and 1970s is a good place to investigate this possibility. Movements of the 1960s were affected by the new technologies that made publication cheaper, easier, and more participatory. These technologies, particularly offset printing, afforded new practices of publication: collaboration, work by amateurs, quick and easy reproduction of images. None of the features of offset technology determined the vernacular style of the counterculture, but popular movements fashioned the features into affordances that supported their colloquial style, informal layout, and extravagant use of images. Genres that deployed and redeployed those elements included participant journalism sponsored by the underground press, the power structure research report developed in the civil rights movement, and adaptations of power structure research by student movements. Women further transformed these practices and genres during the 1970s.

## Affordances of Offset Printing

These techniques of publication were not particularly skilled, and they could take place in convivial, almost recreational, settings. And so a feature of the technology—relative ease of use—became an affordance for the practice of sociability. This affordance was especially marked in the practices of underground papers. David Waddington, a former staff member on the Austin *Rag*, a weekly underground newspaper published in Austin, Texas, from 1966 to 1977, described the paper's layout night as a long party: "Long, long hours on Saturday nights doing layout. Eggs and pancakes at Uncle Van's and the Plantation. . . . Excitement happened whenever Jim Franklin crawled in through the basement window, redolent of patchouli and herbs, down into the Rag Office with the Vulcan ad or perhaps a cover or centerfold. Finally the nights ended with a trip downtown to the Bus Station to put the layout sheets on a bus to Seguin or Waco, trusting the printer to do his job." Alternative newspaper staff members—and anyone who walked in the door could be a staff member—routinely "started in layout" and graduated later to writing and editing. Although layout could be fussy, emphasizing precision and cleanliness, in the alternative press it was lubricated with music and food and emphasized improvisation and fluidity over straight lines and neat corners. Everyone could weigh in on last-minute changes or help choose pictures.

Such a casual approach was possible because the photo-offset press transformed printing from a skilled craft to a routine chore. Someone doing layout in 1962 could have made copies only on a Photostat, a huge, expensive machine that made reverse images on special paper. By 1963, the Xerox 813, a stand-alone plain paper copier, made it possible to produce a quick, cheap photocopy that could be turned into paper printing plates and printed by compact new photo-offset presses. Because the new presses only became cost-effective at three hundred copies, a group of local activists could own and operate a printing press, producing leaflets, posters, and newspapers for other groups in the area, ending their reliance on job printers who might censor a publication if they considered it indecent or unpatriotic (van Uchelen 7). And offset printing was cheap: the first edition of the Austin *Rag*, a thousand copies printed on a photo-offset press, cost sixty dollars; in the history of the paper, its printing bill was never more than two hundred dollars (Olan).

By 1980 it had become clear that photo-offset had changed the printing industry; in a 1981 UNESCO-sponsored book, *Small Printing Houses and Modern Technology*, Roger Jauneau argued that it was no longer worthwhile for developing countries to buy letterpress equipment or to set printed material in hot type: the rotary press, low-cost plates, and photographic composition processes of photo-offset printing were incomparably cheaper and better. He observed that in 1960 letterpress and offset printing had each accounted for 40 percent of the printing done worldwide; photogravure accounted for 20 percent. By 1979 offset accounted for 60 percent of all printing; photogravure and letterpress each accounted for 20 percent (13). When Jauneau suggested that offset printing was the technology of choice for the printing industries of developing countries, he was tracing the affordances of the technology in emerging economies. For the movements of the 1960s, which faced chronic shortages of money, the cheapness and ease of offset printing were critical. But these affordances also translated into social practices of accessibility and conviviality. The technology of offset printing did not determine the practices of the alternative press: in another setting, another culture, cheapness and ease could have afforded a relegation of layout and printing to the lowest levels of a rigid hierarchy. Because the alternative press valued spontaneity and experimentation, they developed differently.

We can get a handle on how offset printing changed layout and pasteup by seeing how these skills were taught in high school print shops. Ralph Maurello's 1960 textbook *How to Do Paste-Ups and Mechanicals* assumed that pasteup would be done for letterpress, that it would be a full-time job, and that it would take a year to learn. Maurello explained, "The work of the paste-up artist necessitates accuracy, precision and neatness. The tools and equipment are simple and few, but must be of good quality, carefully and properly used" (16).

Photographs and other graphics could be included, but only by engraving a separate plate for each image, a process that required both time and money. But when Rod van Uchelen wrote *Paste-Up: Production Techniques and New Applications* sixteen years later, readers needed a much more modest array of tools. Van Uchelen observed that most of this equipment "except for press-type and rubber cement" could be found in any businessperson's desk (14).

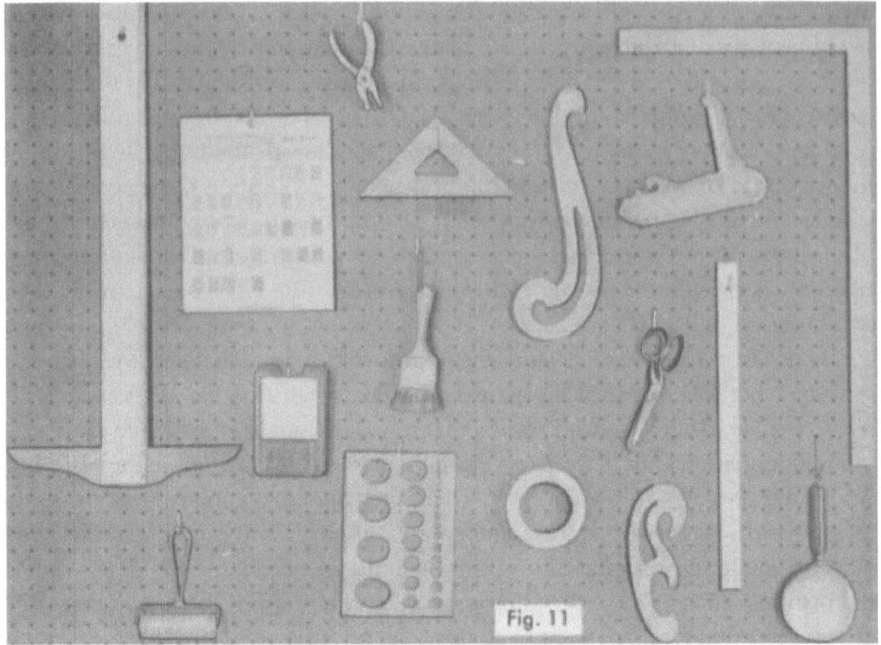

Fig. 8.1 Tools of the trade, 1960. From S. Ralph Maurello, *How to Do Paste-Ups and Mechanicals* (New York: Tudor Publishing, 1960), 43

Van Uchelen's reader did pasteup as a sideline, a routine part of the job; although it was still careful work, pasteup was no longer a skilled, specialized craft, and press-on letters and borders could compensate for uncertain skills with the pen. Because the photographic platemaker did not discriminate among images, hand drawings, lettering, and typescript, the pasteup artist could spontaneously use materials that came to hand and incorporate photographs into the layout without prior planning.

The modest skills van Uchelen described sufficed for laying out an underground paper. These weekly or biweekly local tabloids were central cultural forums for the emerging movements of the 1960s. *Notes from the New Underground*, an anthology of articles from the underground press edited by Jesse Kornbluth and published in 1968, includes such well-known writers as Michael McClure and Tom Robbins (significantly, almost all the writers are male). Kornbluth reports that the Underground Press Syndicate began in 1966 with twenty-five papers, and quickly climbed to fifty. By 1968 the group included a hundred papers (xiv). Abe Peck, in his history of the underground press movement, *Uncovering the Sixties*, reports that in 1971 "nobody knew how many papers were publishing now: eight hundred with ten million U.S. readers was one estimate, four hundred with twenty million was another" (267). There were women's

Technology, Genre, and Gender   155

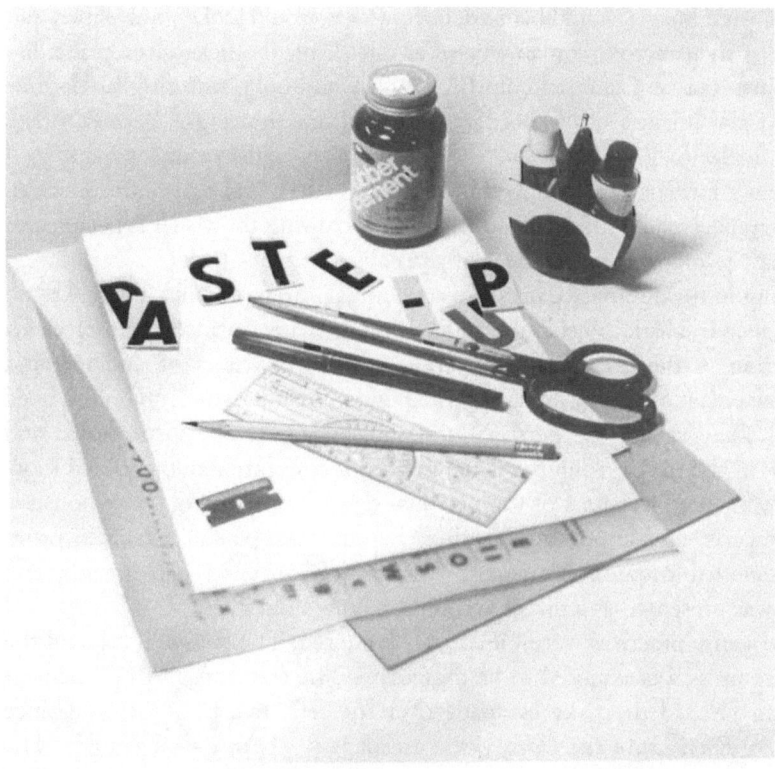

1-1. P-A-S-T-E—U-P spells out the job, and this is the simple equipment required for office use. You can probably find all of it in the average desk, except for the transfer type and the rubber cement.

Fig. 8.2  Tools of the trade, 1976. From Rod van Vchelen, *Paste-Up: Production Techniques and New Applications* (New York: Van Nostrand Reinhold, 1976), 5

papers, GI papers, and many, many high school papers. Most undergrounds stopped publishing or became local advertising sheets before 1980, but at the height of the antiwar movement these papers conducted investigative journalism, cultural reporting, and, through the Liberation News Service, sponsored stringers in Africa, Latin America, and the Middle East.

These papers exploited the affordances of offset printing to develop vernacular printing practices that connected them to volunteer journalists and potential readers. Because the paper was often sold on the street (an archaic distribution practice, to be sure), it needed a striking cover. The undergrounds could draw on some cognate genres for models of attractive images—posters produced in France during the 1968 events, or the ubiquitous concert flyers. Offset technology allowed these images to be reproduced quickly and cheaply. But technology did not determine how the affordance of images would be deployed, what images would be used, or how they would look. Some papers ran a picture of an attractive woman whenever sales lagged; others developed a distinctive layout style: fluid and colorful, or boxy and bold, featuring black-and-white text and images, heavy borders, and bold line drawings.[1] Because offset printing made images cheap and easy to produce, underground papers used them as an affordance for connecting to their readers, and so an affordance of technology became an affordance of genre, borrowing the visual repertoire of the poster to support new practices of journalism.

Writing in the alternative press was colloquial, often profane. There was no attempt at journalistic objectivity; indeed, these writers prided themselves on participating in the events they covered. The core writers of underground papers saw one another constantly, worked under great pressure, and often lived together. They valued the personal and subjective over the institutional and objective; in the undergrounds, all the genres of conventional journalism morphed into so many versions of the personal essay. News stories on demonstrations, concerts, and important meetings became first-person accounts; other stories included ample commentary. Information on car repair, health, and cooking was presented as a direct account of "how I did it."

These genre practices were, of course, gendered. The personal voice of the alternative press was assumed to be masculine. The conventions of participant journalism favored the risky escapade over the reflective response, and since *Huckleberry Finn*, if not the *Odyssey*, the escapade has been coded as male. The work environment of the undergrounds could be unfriendly to women: one of the landmark actions of early feminism was the women's takeover of the New York *Rat* in 1970 after the paper published a sex and pornography issue. Such disputes between men and women were common in the alternative press, but women continued to work on these papers, where they learned how to edit, lay out pages, find advertising, and manage distribution. Offset plates were set on a machine that looked a lot like a typewriter, and typing was a paradigmatic feminine skill. It lacked the masculine whiff of hot lead, and so the layout and printing work of the alternative press was often consigned to women. As the women's movement developed, these skills were put to use, and feminist alternative papers, chapbooks, newsletters, and literary magazines flourished, sometimes published by women's presses (Flannery).

The printed word was no longer the property of experts and skilled tradesmen, but available to anyone; the news was no longer sought out, consumed, or rejected, but produced close to home.

### New Affordances, New Genres: The Power Structure Report

The movements of the 1960s also developed their own genres, exploiting the possibilities of new technologies and refunctioning traditional forms. Kathryn Flannery has written rich accounts of some of these genre practices; one of the

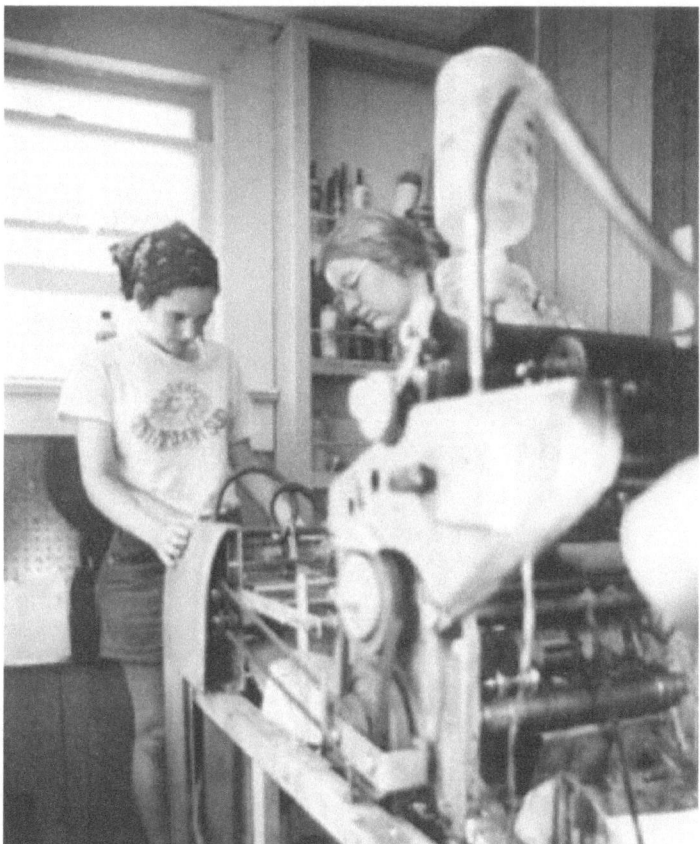

Fig. 8.3 Red River Women's Press, Austin 1973. Photograph courtesy of Danny N. Schweers, www.w2mw.com

most fascinating was the "cranky" (148–56), a simple roll of paper on which a series of images was drawn. The paper was mounted on a mobile frame and could be used either for street performance or more controlled indoor events. The cranky offered its own affordances: it could be set up in a public place in a few minutes, gather a sizable crowd, and, if police intervened, taken away just as quickly. If it were confiscated, vandalized, or abandoned, it was no great loss. These affordances of genre suggested new performance styles: a cranky required fewer people, and less rehearsal, than street theater. At a rally or demonstration, it was a welcome relief from the procession of rabble-rousing speeches.

The power structure research report was a more textually complex genre. Power structure research was central to the curriculum of the 1964 Mississippi Freedom Schools, which taught students how to investigate their local ruling class (Student Nonviolent). Civics was a central subject in the Freedom Schools, and power structure research was a central practice for learning civics. Students

Fig. 8.4 Writing at the Freedom School, 1964. Photograph courtesy of Herbert Randall, Freedom Summer Photographs, Mississippi Digital Library, University of Southern Mississippi

(young people of high school age) were guided by teachers, often volunteer college students, in producing their own local power structure research reports, working under the serious and constant threat of violence. They produced simple documents recording their research and disseminated them locally. Because the civil rights movement was a social laboratory of incomparable power, forms that developed there, including power structure research, were important political resources for all of the insurgent movements of the sixties and seventies.

Members of Students for a Democratic Society (SDS) learned power structure research during their work in Freedom Summer; accounts of their experience circulated in the Economic Research and Action Project (ERAP), an attempt to organize northern and midwestern working-class neighborhoods (Student Nonviolent 2). SDS and ERAP veterans working in student organizations adapted domestic power structure research to make connections between their academic departments and U.S. foreign policy (Schechter; Shapiro).

Fred Goff, a member of the North American Congress on Latin America (NACLA), had been inspired by an SDS pamphlet on the sugar industry and wanted to replicate that research. At the founding meeting of NACLA, in

November 1966, Goff and other activist researchers were mandated to set up a New York office and began publishing the monthly NACLA *Report*. Goff recalls NACLA's response to the student occupation of Columbia: "During the occupation of Columbia University in 1968, we virtually closed down for a few days. NACLA people spent most of their time up there talking to people in the buildings and trying to figure out a more immediate way we could use our research ability. Out of that came the pamphlet, *Who Rules Columbia?* That pamphlet sold a thousand copies the first day" (Shapiro 48). Modeled on both the vernacular model of power structure research and William Domhoff's popular book *Who Rules America?* (1967), the NACLA pamphlet was quickly adapted and disseminated. In style, format, and production, it set the tone for other power structure research pamphlets: printed quickly on newsprint and bound into a letter-sized pamphlet, *Who Rules Columbia?* made liberal use of press-on borders and headlines, did not right justify columns, and offered readers lots and lots of text.

The most striking graphic element in the pamphlet, and in others like it, was the "power structure chart," which adapted the conventions of the corporate organizational chart to demonstrate conflicts of interest, unacknowledged ties, and unsavory connections. The civil rights movement had produced power structure research in a variety of media, from mimeo text to printed booklets, but student activists invariably produced their reports by photo-offset printing. Access to even the minimal technologies of photo-offset was sporadic for civil rights activists, and in any case they used power structure research as much to form the identities of participants as to produce persuasive documents. New Left activists were addressing an audience accustomed to forming their opinions on the basis of printed texts, and they had more access to print technologies. Many power structure research reports were intended to influence student and public opinion during a strike, building takeover, or other kairotic movement; these writers needed both the speed and volume of offset reproduction.

The Columbia document also established the textual features of the genre. The pamphlet supported the demands of the student strike by detailing university collaboration with government policies through its international studies programs and by analyzing Columbia's plans to build a new gym in Harlem. It drew on such public sources as the *New York Times* and corporation prospectuses, but the centerpiece of the book was its reproduction of documents seized by students during their occupation of the university president's office. Again, photo-offset's ability to reproduce an image—in this case, the document itself—became an affordance, enabling readers' direct access to evidence. *Who Rules Columbia?* supported its central claim with swathes of detailed text, all of which were intended to demonstrate that the investment structures of the university took precedence over its educational mission. Detailed information was a sign of writerly authority: a discussion of real estate holdings listed scores of apartment buildings owned by a Columbia trustee, only some of which were relevant to a

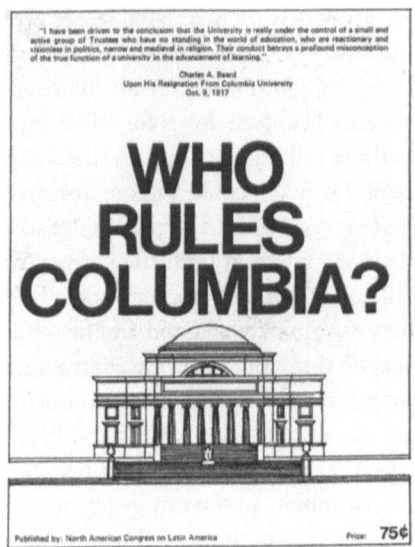

Fig. 8.5 *Who Rules Columbia?*, cover. Courtesy of North American Congress on Latin America

contested development in Morningside Heights and Harlem. The power structure report in its academic setting incorporated norms of comprehensiveness, documentation, and detail that approximated those of scholarly genres.

At least sixteen studies in this genre were published between 1967 and 1975, including *Who Rules Israel?*, *Who Rules the Police?*, and *Who Rules the A.P.A?: A Study of the Backgrounds of Leaders of the American Psychological Association* (Salpeter; Ruchelman; Wood). In 1969, during a student strike, Cambridge activists from ARG (Africa Research Group) and *Old Mole: A Radical Biweekly* collaborated on power structure research focused on Harvard University; in Cambridge, boundaries between the academic left and the alternative press were exceptionally porous. In eight days, working "under great time constraints as well as political and emotional pressures" (Schechter 43), the writers produced an eighty-eight-page booklet, letter-sized, not right justified, and ornamented with a lovely calligraphed power structure chart and a sheaf of memos from the president's office. The report, *How Harvard Rules*, mediated between the psychedelic format of the underground paper and the staid layout of an academic journal: it promised information with an attitude.

The initial readers of *How Harvard Rules* did not quite know what to make of that promise. The Harvard *Crimson* remarked, "What is most fascinating about the book, magazine, or whatever is the range of its analysis." Although they were put off by wooden writing in the pamphlet, the *Crimson* reviewers had to admit that, on some level, "it all holds together." The *Crimson* readers were like participants in a usability study of an unfamiliar technology, struggling to find cognate experiences, searching for the way to interpret features of the text as affordances they could use. Or we could see them as the readers of a new genre, trying to map the new form onto familiar patterns. What they found was a wealth of detail that reinterpreted the dense and particular experience of being at Harvard as a manifestation of the university's network of complicity with the worst policies of the US establishment. Even though the *Crimson* editors might have disagreed with the argument of *How Harvard Rules*, they could not dismiss the new experience of seeing their social context so comprehensively reinterpreted, in real time, under the kairotic pressure of the strike. The technologies of offset printing offered possibilities of production to the writers of power structure research reports; writers realized that those possibilities acted as affordances of the genre, foregrounding some capabilities (reaching large audiences) and muting others (training inexperienced researchers).

### Affordances of Genre?

How, then, did technology and genre interact in the 1960s and 1970s? In describing both the simple cranky and the developed power structure research report, I referred to "affordances of genre," and it is not unusual to discuss genre in these terms, as if it were an environment or a technology. For example, in his genre analysis of blogging, Lucas Graves observes, "In some sense, a genre is a set of affordances, the communicative template that results when culture renders technological possibility" (338). For many genre theorists, the concept of affordance is linked with the metaphor of technology as a text (Hutchby, "Technologies"; Oliver)—a technology is seen as requiring interpretation and performance, like a book or a play. When the technology, understood as a text, produces a text characterized by its own affordances, technology and text become, reciprocally, metaphors for each other. The ease of doing layout for offset printing supports the convivial publication of underground papers; those papers adopt formal features—styles of writing and genre preferences—that express practices of sociability and amateur production among its readers. The cranky facilitated quick, impromptu performance; feminist groups assimilated these affordances and constructed a performance genre of "zap actions," including spray paint graffiti on offensive posters and the distribution of stickers reading "This offends women." The zap action transformed capacities of technologies, new and old, into affordances for group activity and expressive action. (Spray paint had been in distribution only since the mid-1950s, but gummed stickers had been available since the nineteenth century.) These technologies were coded as affordances of speed and adapted to a genre whose very name—the zap action—invoked a short, spontaneous performance. The zap action afforded quick, imaginative collective action and also, as an alternative to New Left organizational forms, presented an argument about how collective consciousness was formed.

## Power Structure Research: Feminist Appropriations and Adaptations

There is a single instance that I know of in which feminists appropriated the power structure research report. This report torqued the emerging genre. *How Harvard Rules Women*, produced by women of the New University Conference, a college-based New Left organization, was published during the same student strike as was *How Harvard Rules*; both reports were excerpted in the *Old Mole*. *How Harvard Rules Women* followed the emerging format conventions of power structure research: letter-sized newsprint pages, a full-sized image on the cover, unjustified print, and oceans of unbroken text.

But in place of the power structure chart and captured memos, *How Harvard Rules Women* offered scathing accounts of routine discrimination against women students, faculty, and staff. Power structure research had offered these women a set of genre affordances: a paradoxical combination of relentless focus and endlessly exfoliating details; an interest in secrets and their revelation; a gesture of unveiling the political consequences of mundane practices. Women took up these affordances and reshaped the genre with personal, narrative discourses

Fig. 8.6 *How Harvard Rules Women*, cover. Courtesy of New University Conference

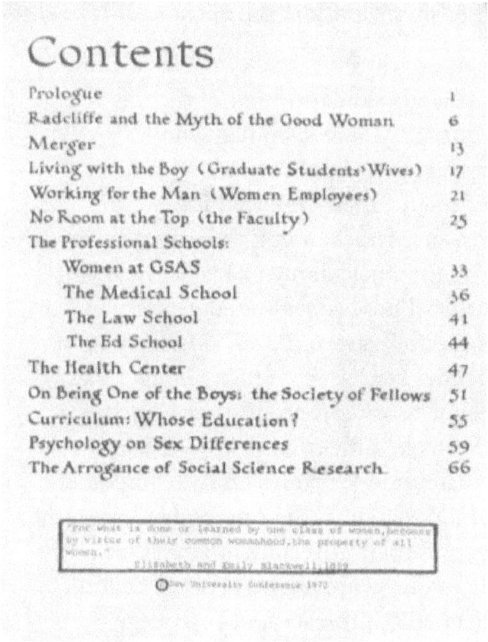

## Contents

Prologue ................................................................. 1
Radcliffe and the Myth of the Good Woman ............ 6
Merger .................................................................. 13
Living with the Boy (Graduate Students' Wives) ...... 17
Working for the Man (Women Employees) .............. 21
No Room at the Top (the Faculty) .......................... 25
The Professional Schools:
    Women at GSAS ................................................. 33
    The Medical School ............................................ 36
    The Law School .................................................. 41
    The Ed School .................................................... 44
The Health Center ................................................ 47
On Being One of the Boys: the Society of Fellows ... 51
Curriculum: Whose Education? .............................. 55
Psychology on Sex Differences .............................. 59
The Arrogance of Social Science Research ............. 66

> "For what is done or learned by one class of women, becomes by virtue of their common womanhood, the property of all women."
>
>             — Elizabeth and Emily Blackwell, 1859
>
>         © New University Conference 1970

Fig. 8.7 *How Harvard Rules Women*, table of contents. Courtesy of New University Conference

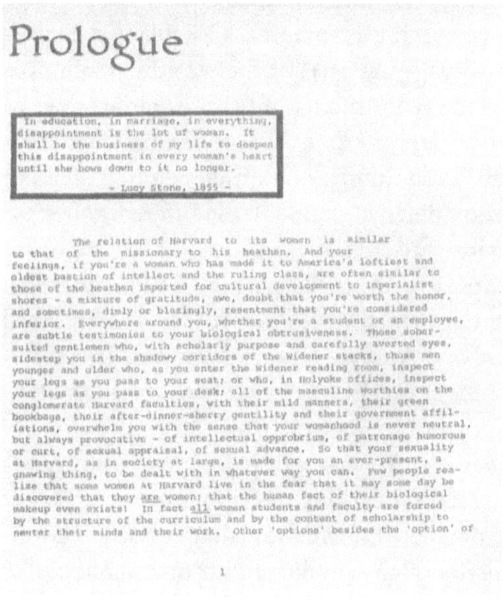

## Prologue

> "In education, in marriage, in everything, disappointment is the lot of woman. It shall be the business of my life to deepen this disappointment in every woman's heart until she bows down to it no longer."
>
>       — Lucy Stone, 1855 —

    The relation of Harvard to its women is similar to that of the missionary to his heathen. And your feelings, if you're a woman who has made it to America's loftiest and oldest bastion of intellect and the ruling class, are often similar to those of the heathen imported for cultural development to imperialist shores — a mixture of gratitude, awe, doubt that you're worth the honor, and sometimes, dimly or blazingly, resentment that you're considered inferior. Everywhere around you, whether you're a student or an employee, are subtle testimonies to your biological obtrusiveness. Those sober-suited gentlemen who, with scholarly purpose and carefully averted eyes, sidestep you in the shadowy corridors of the Widener stacks, those men younger and older who, as you enter the Widener reading room, inspect your legs as you pass to your seat; or who, in Holyoke offices, inspect your legs as you pass to your desk; all of the masculine Worthies on the conglomerate Harvard faculties, with their mild manners, their green bookbags, their after-dinner-sherry gentility and their government affiliations, overwhelm you with the sense that your womanhood is never neutral, but always provocative — of intellectual opprobrium, of patronage humorous or curt, of sexual appraisal, of sexual advance. So that your sexuality at Harvard, as in society at large, is made for you an ever-present, a gnawing thing, to be dealt with in whatever way you can. Few people realize that some women at Harvard live in the fear that it may some day be discovered that they are women; that the human fact of their biological makeup even exists! In fact all women students and faculty are forced by the structure of the curriculum and by the content of scholarship to neuter their minds and their work. Other 'options' besides the 'option' of

1

Fig. 8.8 *How Harvard Rules Women*, first page. Courtesy of New University Conference

that were emerging in consciousness-raising. Consider the opening of *How Harvard Rules Women*.

> The relation of Harvard to its women is similar to that of the missionary to his heathen. And your feelings, if you're a woman who has made it to America's loftiest and oldest bastion of intellect and the ruling class, are often similar to those of the heathen imported for cultural development to imperialist shores—a mixture of gratitude, awe, doubt that you're worth the honor, and sometimes, dimly or blazingly, resentment that you're considered inferior. Those sober-suited gentlemen who, with scholarly purpose and carefully averted eyes, sidestep you in the shadowy corridors of the Widener stacks, those men younger and older who, as you enter the Widener reading room inspect your legs as you pass to your set; or who, in Holyoke offices, inspect your legs as you pass to your desk; all of the masculine Worthies on the conglomerate Harvard faculties, with their mild manners, their green bookbags, their after-dinner-sherry gentility and their government affiliations, overwhelm you with the sense that your womanhood is never neutral, but always provocative—of intellectual opprobrium, of patronage humorous or curt, of sexual appraisal, of sexual advance. (1)

The exigency of power structure research was to unmask established institutions: democracy in Mississippi was actually the rule of the wealthy; Columbia University made decisions to protect the investments of trustees rather than to improve the education of students. *How Harvard Rules Women* demonstrated that Harvard was not a rarefied intellectual community, but a men's club. Instead of detailed lists of real estate holdings or defense contracts, it offered countless examples of daily humiliation. Intrinsic to the message of power structure research was the performance of exposure by those who had been invisible: African American farmers, or graduate students, or women at Harvard. Harvard women, Columbia graduate students, and African American farmers in Mississippi had, of course, very little in common, except that nobody expected them to speak so eloquently about the conditions of their lives.

Besides *How Harvard Rules Women* there were, as far as I know, no other instances of feminist power structure research. But the affordances of power structure research—presentation of detailed information that revised conventional wisdom; research by lay members of the public; and broad publicity for formerly restricted information—were transposed into another feminist project. When the members of the group that would become the Boston Women's Health Book Collective decided, with some trepidation, to publish the notes for a course they had taught, the format of the power structure research book and the specific example of *How Harvard Rules Women* offered them a model for a modest, participatory, and heavily researched pamphlet. *Women and Their Bodies*

(1970), the book they produced, bears a family resemblance to *How Harvard Rules Women;* both texts rely heavily on the capacities of photo offset printing as they were realized in the genre conventions of power structure research; they mobilize the affordances that ten years of vernacular publishing had developed. The covers of both *How Harvard Rules Women* and *Women and Their Bodies* are illustrated with a single photograph and a lettered title; both books sold for seventy-five cents; both are printed on newsprint; both have hand-drawn tables of contents.

Both books demonstrated, in their material features and the texture of their writing, that publication, like the performance of music, could become something that groups of friends undertook as a project: a quick convivial movement from the typewriter to the printed page, rather than a solitary, multiyear, life-defining project.

*Women and Their Bodies* was the first women's health manual written by ordinary lay women, and it was very different from the previously published general family medical references, "baby books," or "marriage manuals." The look and feel of the book invoked vernacular publishing practices and established its relation to readers: this was not a patronizing book written by a doctor; it had been put together by "ordinary women" offering advice based on their experiences. The text solicited readers to do their own investigations, and the material form of the book assured them that they could very well do their own publication, too.

Fig. 8.9 *Women and Their Bodies*, 1970, cover. Courtesy of Boston Women's Health Book Collective

TABLE of CONTENTS



THIS PAPER HAS BEEN LAID OUT SO THAT IT MAY BE USED EITHER AS IT IS – IN 4 BOUND BOOKLETS OR AS SEPARATE SHEETS IN A RING BINDER. FOR A NOTEBOOK: PUNCH HOLES IN THE WIDE MARGINS AND SLIT THE BINDING THREAD AND THE BACK OF EACH BOOKLET WITH A RAZOR BLADE



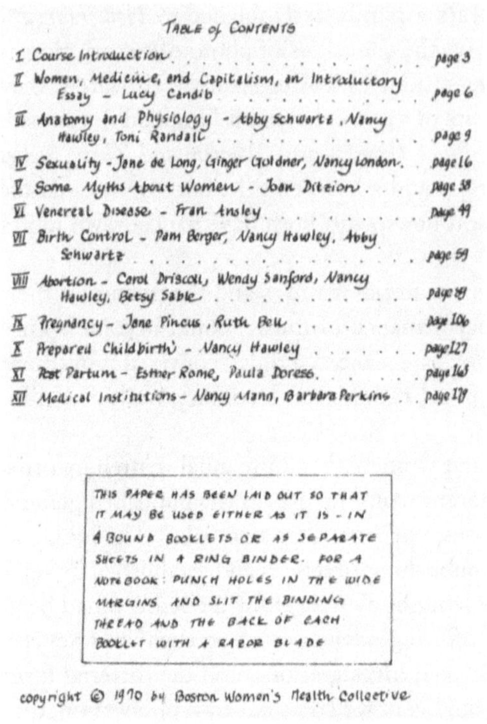

Fig. 8.10 *Women and Their Bodies*, 1970, table of contents. Courtesy of Boston Women's Health Book Collective

3

COURSE INTRODUCTION

One year ago, a group of us who were then in women's liberation (now most of us consider ourselves members of Bread and Roses) got together to work on a laywoman's course on health, women and our bodies. The impetus for this course grew out of a workshop on "women and their bodies" at a women's conference at Emmanuel College in Boston, May 1969. After that, several of us developed a questionnaire about women's feelings about their bodies and their relationship to doctors. We discovered there were no "good" doctors and we had to learn for ourselves. We talked about our own experiences and we shared our own knowledge. We went to books and to medically trained people for more information. We decided on the topics collectively. (Originally, they included: Patient as Victim; Sexuality; Anatomy; Birth Control; Abortion; Pregnancy; Prepared Childbirth; Postpartum and Childcare; Medical Institutions; Medical Laws; and Organizing for Change.) We picked the one or ones we wanted to do and worked individually and in groups to write the papers. The process that developed in the group became as important as the material we were learning. For the first time, we were doing research and writing papers that were about us and for us. We were excited and our excitement was powerful. We wanted to share both the excitement and the material we were learning with our sisters. We saw ourselves differently and our lives began to change.

As we worked, we met weekly to discuss what we were learning about ourselves, our bodies, health and women. We presented each topic to the group, gave support and helpful criticisms to each other and rewrote the papers. By the fall, we were ready to share our collective knowledge with other sisters. Excited and nervous (we were just women; what authority did we have in matters of medicine and health?), we offered a course to sisters in women's liberation. Singularly and in groups, we presented the topics and discussed the material; sometimes in one large group, often in smaller groups. Sisters added their experiences, questions, fears, feelings, excitement. It was dynamic! We all learned together.

One original version of the course was that we as a group would give the course to a group of women who could then go out and give it to other women. To some extent, that is what happened. After the first time around, those of us who had worked out the course originally, plus women who had taken the course, got together in an enlarged group to rewrite the papers so they could be printed and shared, not only with women in Boston, but with women across the country. Other women wanted to learn. Other women's health groups wanted to compare and combine our work and theirs.

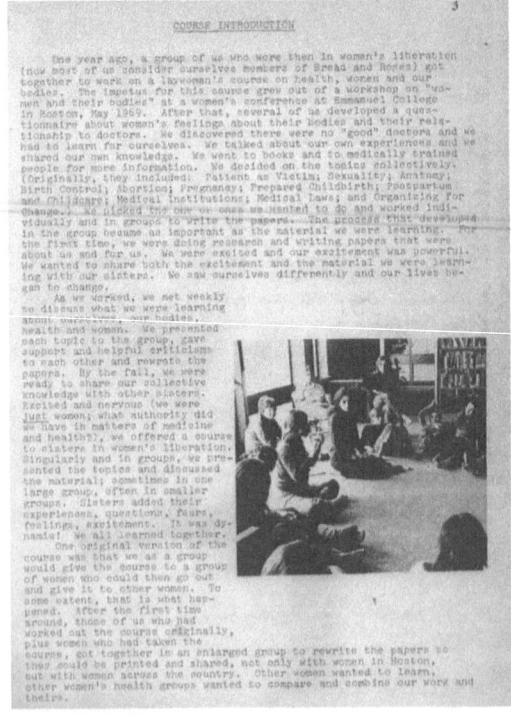

Fig. 8.11 *Women and Their Bodies*, 1970, first page. Courtesy of Boston Women's Health Book Collective

The affordances of the text rhymed with the affordances of the technology that produced it. *Women and Their Bodies* was wildly popular; the title was changed to *Our Bodies, Our Selves* in 1971, and again to *Our Bodies, Ourselves* when the book was published by Simon and Schuster in 1973. Printed on newsprint for $1,500 (Hawley), it was bound, as the note inside the front cover said, "so that it may be used either as it is—in four bound booklets or as separate sheets in a ring binder." Readers were given instructions for taking the book apart and reassembling it. Not only did *Women and Their Bodies* look homemade; it invited readers to remake it in their own homes.

Unlike the spartan graphic presentation of the power structure research report, *Our Bodies, Ourselves* exploited the emerging capabilities of photo-offset printing in its use of photographs. The collective did not favor the explosive graphic style, neon colors, or experimentation with format of other movement publications, which used poster-size paper and flurries of typewritten blurbs, or juxtaposed graphics with poetry. *Our Bodies, Ourselves* was laid out quite conventionally, with text and graphics aligned to the edge of the page. What was new in *Our Bodies, Ourselves* was its use of images: the book included photographs of the writers, their friends, and their families. An image showing a collective member standing naked over a mirror, labeled "Esther's vulva" in the archived copy, illustrated the anatomy chapter. As members of the Boston Women's Health Book Collective transposed power structure research into a new genre, they used offset printing's capacity for reproducing images as an affordance for reimagining how women's bodies could be presented. The collective commissioned line drawings to illustrate the book, integrating anatomical information with images of women who had faces, who gestured, who moved.

*Our Bodies, Ourselves*, like the power structure research report, constructed ethos by offering detailed information rather than by claiming academic authority. Instead of "liberating" memos, writers from the collective sneaked into Harvard's Countway Medical Library, "liberating" medical information from closed professional circulation. (The medical library at Boston University was open to the public but did not offer the thrill of transgressive entry into the Countway, open only to students and faculty of the Harvard Medical School.) They described their favorite methods of masturbation, their excitement at the discovery of the clitoris, their worries about birth control pills. Just as power structure research encouraged readers to produce their own pamphlets, *Our Bodies, Ourselves* encouraged readers to investigate their own bodies, draw their own conclusions, and do their own publication.

### Affordance and Identity

The affordances of technology reciprocated the affordances of genre, supporting practices of publication that made power structure research reports, and the texts related to them, consequential for both readers and writers. These books were material evidence of actions their writers had undertaken—participating in Freedom Schools, occupying the university president's office, raiding the medical library. They were intended to provoke actions among their readers—registering to vote, abstaining from military research, questioning one's doctor. Earlier in this essay, I suggested that power structure research projects were valued for both the documents they produced and the styles of interaction they fostered. And such a dual function is often characteristic of situated genres: Catherine Schryer and her collaborators demonstrate how, in the medical case presentation, genre performance works at once to produce knowledge and to develop professional identities. The case presentation communicates information about patient care and also cultivates a student's sense of a particular profession to which she aspires.

The organizers of the Freedom Schools saw power structure research as a tactic for interrupting monolithic white supremacy, for producing political knowledge that supported a new civic identity. James Silver, president of the Southern Historical Association, described the culture of Mississippi in 1963 as based on "a never ceasing propagation of the 'true faith' [of white supremacy]," enforced with a "constantly reiterated demand for loyalty to the united front demanding that nonconformists be hushed, silenced with a vengeance, or in crisis situations driven from the community. Violence and the threat of violence have reinforced the presumption of unanimity" (3–4). Against this violently enforced "presumption of unanimity," the Freedom School organizers raised issues about students' lives to "stimulate latent talents and interests that have been submerged too long . . . causing high school youth in Mississippi to QUESTION" (qtd. in Perlstein 309). For them, power structure research produced a useful document, but it also produced writers who would question received ideology.

The power structure reports of the student movement supported particular political demands but also fostered a skeptical, dissenting civic identity. Writers of *How Harvard Rules* anticipated that their readers would be skeptical of the foldout power structure chart included in the text and invited them to do their own research: "By now some eyes will be blinking in disbelief. Can this be true, they will ask. It appears so overdone! It smacks of a crude conspiracy theory of power. . . . If we still haven't told you enough, don't despair. Pick up the *who's who, social register* and *moody's manual*. Then, make your own chart. It will do funny things to your head too."

For the writers of *Our Bodies, Ourselves*, the health book (or, originally, the health course) had a similar double function. It would give women information that they needed, and it would confirm women's trust in their own experience of embodiment and the support of other women. Readers were invited to begin new conversations: "It was exciting to learn new facts about our bodies, but it was even more exciting to talk about how we felt about our bodies, how we felt about ourselves, how we could act together on our collective knowledge to change the health care system for women and for all people. We hope this will be true for you, too" (Boston Women's Health 4). These texts were all seen as affordances supporting a certain style of political organization: decentralized, spontaneous, and populist. Activists would become their own experts and publicists, their own publishers and distributors. It was a romantic vision, and it had its limits. The Boston Women's Health Book Collective's desire to provide readers with comprehensive, reliable health information was at odds with their commitment to an egalitarian, nonexpert practice of knowledge. As early as 1973, group members mourned that "we never talk anymore," or "we don't have real, close conversations like we used to" ("Minutes"). The affordances of power structure research did not support sustained investigation or growing expertise.

Produced at moments of political crisis, power structure research reports expressed a collective intellectual identity that hybridized professional research skills and vernacular publication practices. That hybrid identity was fragile: some groups, such as NACLA and Health-PAC, developed a line of research that was supported by foundations and donors and became professionalized. Others resolved themselves into the counterculture: some writers of *Who Rules Harvard?* continued to work as professional academics, coordinating their scholarship through the Africa Research Group. Still others continued to write for *Old Mole: A Radical Biweekly* and to relate to the paper's countercultural base (Albert 113). In all the examples I have presented, the genre of power structure research could function as either a means of identity formation or as a means for presenting and developing information, but it could not serve both purposes in any sustained way: the more skilled a writer became at the task of exposure, the less exemplary was her work. Once a writer had worked on a power structure research report, that writer was no longer the ideal author of a power structure research report. This interference between genre as a way of forming identity and genre as a way of organizing texts was not negotiable: we have now a literature of exposure and muckraking (such as *Fast Food Nation*), but no practice of vernacular research, except, perhaps, as it may be developing in community writing programs.

These publication practices raise questions about contemporary technologies of writing, which also offer the promise of relatively democratic access and amateur production. Do vernacular digital media, unlike the alternative publications of the 1960s, have affordances that will sustain them after an initial flush of enthusiasm? How are the affordances emerging with digital genres being torqued and transformed by the work of formerly excluded groups? Is the stability of an "institution" like *Our Bodies, Ourselves* something that contemporary practitioners want to emulate? If so, what are the best strategies for doing so?

The history of these publication practices and genres also suggests how rhetoricians might think about new media and their affordances. This history richly demonstrates that, although different media and genres offer different affordances, material affordances do not determine how writers and readers will deploy technologies or genres. Offset printing enables writers to include images in their text; nobody would have anticipated that "Esther's vulva" would have been among them. "New media" is a constantly mobile term: there are always new media, and they are always suggesting new practices for producing and disseminating texts. New technical resources, the practices they foster, and the forms they suggest become elements of the kairos that prompts the invention of new genres: the occasion that gives rise to a rhetorical performance includes the means by which that performance is organized and disseminated. Like the new digital genres Carolyn Miller and Dawn Shepherd have described, power structure research arose "from a dynamic, adaptive relationship between discourse and kairos." The alternative newspaper and the power structure report did serious rhetorical work: they organized rhetorical resources and supported ongoing social movements. Neither form survived the decline of these movements and the dispersal of their resources. The cognate genres developed by the women's movement—the health book, health narratives, personal stories of transformation—have survived the decline of the movements that originated them, but they have become affordances responding to a new set of exigencies, and now reorganize affective life under the new conditions that women face. It is an open question, and not an easy one, whether the affordances of these forms might preserve certain discursive energies after the occasions that excited them have passed. Whether those energies can (or should) be recovered for political life is another, even more difficult question.

## Note

1. For a rich collection of underground newspaper covers, see "Voices from the Underground and Radical Press in the 'Sixties': An Exhibition."

## Works Cited

ARG [Africa Research Group] and *Old Mole*. *How Harvard Rules: Being a Total Critique of Harvard University, Including: New Liberated Documents, Government Research, the Educational Process Exposed, Strike Posters, & a Free Power Chart*. [Cambridge, Mass.: 1969].

Albert, Judith Clavir, and Steward Edward Albert. *The Sixties Papers: Documents of a Rebellious Decade*. New York: Praeger, 1984.

Boston Women's Health Collective. *Women and Their Bodies: A Course*. Boston: Boston Women's Health Collective, 1970.

Cogoli, John. *Photo-Offset Fundamentals*. Bloomington, Ind.: McKnight, 1973.

Domhoff, William. *Who Rules America?* New York: Prentice Hall, 1967.

Flannery, Kathryn Thoms. *Feminist Literacies, 1968–75*. Urbana: University of Illinois Press, 2005.

Gaver, William W. "Affordances for Interaction: The Social Is Material for Design." *Ecological Psychology* 8 (1996): 111–29.

Gibson, James. *An Ecological Approach to Visual Perception.* Boston: Houghton Mifflin, 1979.

Graves, Lucas. "The Affordances of Blogging: A Case Study in Culture and Technological Effects." *Journal of Communication Inquiry* 31 (2007): 331–46.

Hawley, Nancy Miriam. Personal interview. April 2006.

"Herbert Randall Freedom Summer Photographs." 20 December 2008 http://www.lib.usm.edu/~archives/m351.htm?m351text.htm~mainFrame.

"How Harvard Rules." *Harvard Crimson*, 5 July 1969. Accessed 2 August 2008, http://www.thecrimson.com/article.aspx?ref=210032.

*How Harvard Rules Women.* Cambridge, Mass.: New University Conference, 1970.

Hutchby, Ian. "Affordances and the Analysis of Technologically Mediated Interaction: A Response to Brian Rappert." *Sociology* 37 (2003): 581–89.

———. *Conversation and Technology.* London: Polity, 2001.

———. "Technologies, Texts, and Affordances." *Sociology* 35 (2001): 441–56.

Jauneau, Roger. *Small Printing Houses and Modern Technology.* Paris: UNESCO, 1980.

Jewitt, Carey, and Gunther Kress, eds. *Multimodal Literacy.* London: Peter Lang, 2003.

Kornbluth, Jesse, ed. *Notes from the New Underground.* New York: Viking, 1968.

Maurello, S. Ralph. *How to Do Pasteups and Mechanicals: The Preparation of Art for Reproduction.* New York: Tudor, 1960.

Miller, Carolyn, and Dawn Shepherd. "Blogging as Social Action: A Genre Analysis of the Weblog." In *Into the Blogosphere: Rhetoric, Community, and the Culture of Weblogs*, edited by Laura Gurak et al. University of Minnesota Libraries, 2004. Accessed 19 December 2008, http://blog.lib.umn.edu/blogosphere/blogging_as_social_action.html.

"Minutes." Boston Women's Health Book Collective. 26 September 1973. Radcliffe Institute, Harvard University. Schlesinger MC503, carton 4, folder 16.

Olan, Susan. "The *Rag*: A Study in Underground Journalism." Master's thesis, University of Texas at Austin, 1981. Accessed 3 August 2008, http://www.utwatch.org/archives/ragthesis.html#ch1.

Oliver, Martin. "The Problem with Affordance." *E-Learning* 2.4 (2005): 402–13.

Peck, Abe. *Uncovering the Sixties: The Life and Times of the Underground Press.* New York: Pantheon, 1985.

Perlstein, Daniel. "Teaching Freedom: SNCC and the Creation of the Mississippi Freedom Schools." *History of Education Quarterly* 30.3 (1990): 297–324.

Pfaffenberger, Bryan. "Technological Dramas." *Science, Technology and Human Values* 17 (1992): 282–312.

Rappert, Brian. "Technologies, Texts, and Possibilities: A Reply to Hutchins." *Sociology* 37.3 (2003): 565–80.

Ruchelman, Leonard. *Who Rules the Police?* New York: NYU Press, 1973.

Salpeter, Eliahu. *Who Rules Israel?* New York: Harper and Row, 1973.

Schechter, Daniel. "From a Closed Filing Cabinet: The Life and Times of the Africa Research Group." *Issue: A Journal of Opinion* 6.23 (1976): 41–48.

Schryer, Catherine, Lorelei Lingard, Marlee Spafford, and Kim Garwood. "Structure and Agency in Medical Case Presentations." In *Writing Selves / Writing Societies: Research from Activity Perspectives*, edited by Charles Bazerman and David R. Russell.

Fort Collins, Colo.: The WAC Clearinghouse and Mind, Culture, and Activity, 2003. Accessed 3 August 2008, http://wac.colostate.edu/books/selves_societies.

Schweers, Danny. "Photos at *The Rag*: 1971–1975." Accessed 3 August 2008, http://www.w2mw.com/galleries/Rag1970s/index.htm.

Shapiro, Helen. "NACLA Reminiscences: an Oral History." *NACLA Report* 15.5 (1981): 45–56.

Silver, James. "Mississippi: The Closed Society." *Journal of Southern History* 30.1 (1964): 3–34.

Student Nonviolent Co-ordinating Committee [SNCC]. "Freedom School Curriculum." The Student Nonviolent Co-ordinating Committee Papers, 1959–1972. Accessed 19 December 2008, http:www.education and democracy.org/index.html.

van Uchelen, Rod. *Paste-Up: Production Techniques and New Applications*. New York: Van Nostrand Reinhold, 1976.

"Voices from the Underground and Radical Press in the 'Sixties': An Exhibition." 26 October 2005. University of Connecticut Libraries. Accessed 3 August 2008, http://www.lib.uconn.edu/online/research/speclib/ASC/exhibits/voices.

Waddington, Rick. "*Rag* Memoirs: 2005." Accessed 3 August 2008, http://www.accyes.org/RagStaffMemoirs.asp#cam.

*Who Rules Columbia?* New York: NACLA, 1967. Republished at "UtahWatch.org" January 2008. Accessed 3 August 2008, http://www.utwatch.org/archives/whorulescolumbia.html.

Wood, Victor. *Who Rules the A.P.A? A Study of the Backgrounds of Leaders of the American Psychological Association*. Arcata, Calif., 1973.

# Rhetoric in (as) a Digital Economy

James E. Porter

In this essay I explore the theoretical implications of the paradigmatic shift occasioned by the technological developments of Web 2.0,[1] focusing on how the emerging digital economy of Web 2.0 is changing, or ought to change, our notions of rhetoric and writing. A playful subtitle for this essay might be "How Do 'the Long Tail' and 'the Wisdom of the Crowd' Matter to Rhetoric and Writing?" The two phrases refer to two popular books: Chris Anderson's 2006 book *The Long Tail: Why the Future of Business Is Selling Less of More* (based on his 2004 *Wired* magazine article "The Long Tail") and James Surowiecki's 2004 book *The Wisdom of Crowds*. In brief, long-tail economics refers to a key feature of digital economics: because of the low cost of selling and distributing digital information (and even of selling nondigital products via digital means), it is possible to sell products to smaller market niches than in nondigital economies. In rhetorical terms this means that it is economically feasible to design and distribute tailored information for smaller audience groups (versus trying to make a single information product work for a larger general audience).[2] Social networking refers to Web 2.0 applications that are built from user-generated content. Social networking Web sites such as Flickr, Delicious, YouTube, and others coordinate the power of many contributors, many of them amateur contributors, and sift their creations through a system that assesses their value.

Simply put, my overall point here is that developments in network-based technology—particularly the emergence and success of "the networked information economy" (Benkler, *Wealth*) and of Web 2.0 social networking—will dramatically change rhetoric theory and the practice of writing. What I mainly do in this essay is take a scene-act perspective—to use one of the dyads out of Kenneth Burke's dramatistic rhetorical method. In Burkean terms, I am examining how the *scene* of digital writing (particularly the social dynamic of the Internet) affects the *act* of writing and composing. But at the end I shift to a scene-agent perspective that looks at the *who*, particularly the whos that are, or could be, excluded or exploited in this emerging scene.

I begin with a consideration of the relationship between rhetoric, digital economics, and delivery, arguing that economics is, or should be, a key component of any rhetoric theory. I then discuss digital economics and social networking—"the long tail" and "the wisdom of the crowd"—describing these phenomena and pointing to their implications for rhetoric and for writers. (I illustrate this point via discussion of a research project I worked on through the WIDE Research Center at Michigan State University.) I generally applaud this move toward social networking—as it represents empowerment of the audience in rhetorical interactions—but I also consider the darker side of social networking and of economic systems based on user-generated content.

## Rhetoric, Digital Economics, and Delivery

As I have argued elsewhere, economics has always been an important component of rhetoric,[3] but historically the relationship has only occasionally been articulated, appreciated, or examined within the field of rhetoric—most notably by Deirdre McCloskey and Richard Lanham (see also Johnson-Eilola, "Relocating"; Johnson-Eilola, "Accumulation"; Carter; Salvo).

I need to distinguish my treatment from both McCloskey's and Lanham's. My focus is *the economics of rhetoric*, not *the rhetoric of economics*. Whereas McCloskey looks at how rhetoric plays a role in the field of economics, I am looking at the economics of rhetoric—that is, how rhetorical contexts themselves rely on an economic system of exchange. The exchange is not always a commercial one, but there is an exchange of value that serves as the motivation for the production and circulation of rhetorical objects. So, in linking rhetoric and economics, I am not doing it à la McCloskey.

Nor am I doing it quite à la Richard Lanham. In *The Economics of Attention*, Lanham argues that in the digital age we need a new economic model—an economy of attention based on rhetoric, which he sees from a stylistic and design perspective as the art of deploying creative, imaginative, and innovative techniques for grabbing and keeping audience attention. In this realm—and I would agree with Lanham on this point—specific domain expertise matters less, rhetoric matters more. However, Lanham's stylistic view of rhetoric misses an essential point about the digital economy. It is not just about style; it is also about substance and value (see Goldhaber). A broader view would see rhetoric as requiring a productive and pragmatic knowledge about how to create information products that will matter to people—that is, be usable and useful. A broader view of rhetoric would include inquiry procedures (that is, inventional tactics) aimed at understanding what motivates people to create, search, and circulate knowledge.

In other words, the digital economy needs a robust view of rhetoric, a view that includes inventional procedures for developing knowledge and for collaborating with audiences to co-create knowledge.

Classical Roman rhetoric had two terms for the development of content and for the distribution of information products: *inventio* and *actio*. These two concerns—invention and delivery—are two of the three historically neglected canons of rhetoric (memory being the third). That neglect has to do with the persistence in recent history and in popular culture of a predominant alphabetic, print-based view of thinking about writing. The print view sees writing in reductive terms as "words on a page." That view still sees writing instruction as mainly a matter of teaching style and arrangement (syntax and diction, grammatical competency, arrangement of ideas on the page), and teaching it mainly within the realm of print. That is the narrow, instrumental view of writing: writing is simply the words you choose to convey your message and how you organize them on the page. The "content" for your writing comes from someplace else (that is, from real disciplines). Rhetoric as the dress of thought. It is important to note, though, that Lanham has a high degree of respect for the dress of thought. Clothing matters to him; there is a good deal of substance in style, he argues. But nonetheless he holds to a binary view that fundamentally sees content development as something outside the realm of rhetoric. As does Peter Ramus, he disconnects invention from rhetoric.

But there is another view of writing, the substantive view, in which writing has a much larger scope. In the substantive view, the art of writing includes understanding the entire scene or context of communication, inventing and developing content, determining audience needs, constructing effective arguments, designing effective interfaces, compiling evidence, understanding community and cultural values, figuring out where and how to deliver the message (through what technological means), coordinating and collaborating among various writers and groups, predicting the flows and interrelationships among the elements of communication, ad infinitum. In short, writing involves a bunch of decisions, issues, and questions that involve critical thinking, deep analysis of communication situations, and both theoretical and practical how-to knowledge. This set of concerns *is* part of the art of rhetoric—and that is what is most changed in digital environments.

When rhetoric asks questions about audience and purpose—What is my purpose for writing? Who is my audience?—it is also implicitly asking questions about the economics of delivery. What motivates someone to produce and distribute a piece of writing? What motivates someone else to access it, read it, interact with it? What drives the interaction and makes it productive for both parties? These are basic questions of digital economics, but also basic questions for rhetoric, particularly for the canon of delivery (Porter, "Recovering Delivery"; Eyman, "Digital Rhetoric").

Why do we write? The stock answer in rhetoric and composition has often been something like "to inform, to persuade, to entertain." But why would anyone want to inform somebody or create a poem? What is the point of doing *that*? There is another calculus involved in any act of writing: purpose in the sense of *value*. There must be some value for the reader or for the writer in the act of producing, distributing, exchanging texts. Somebody has information; somebody else needs it. Somebody wants to express a feeling; somebody else needs to feel it. But what motivates such an exchange? Writing—*all* writing, I would say—resides in economic systems of value, exchange, and capital. Not necessarily monetary or commercial systems—think about Bourdieu's notions of cultural capital and social capital[4]—but economic systems nonetheless. The kind of economics I am talking about has to do with value more broadly defined. It might well involve the exchange of money, but the motivation could just as easily be based on desire, participation, sharing, emotional connectedness. This is the secret of the Web 2.0 dynamic.

This broader sense of value helps to explain the proliferation of blogs on the Web and the growing number of entries in spaces like Wikipedia. As Clay Shirky has said, from an economic standpoint, "It sure is weird that the Wikipedia works" (qtd. in Aigrain). It is not weird if you accept that people write because they want to interact, to share, to learn, to play, and to help others. That drive of people to interact socially is a key feature of the new digital economy—and the rhetorical basis of social networking.[5] It explains the popularity of blogs and of social networking spaces such as Facebook, MySpace, and YouTube.

### The Digital Economics of "The Long Tail"

"The long tail," a term coined by Chris Anderson, refers to how conventional economic models are overturned by Web-based communications (see Shirky, "Power Laws"). In the digital realm of no-cost reproduction and low-cost distribution, it is economically viable to make money on products that have a low sales volume. Attracting a wide readership (market) is no longer as important, not when you are talking about a product that costs very little to reproduce and distribute. The cost of distribution is so low in the digital realm that I can invest my energy manufacturing a product that sells to only twelve people (or even two). This is a very different economic model from that of the manufacturing economy—or of the print economy—and rhetoric needs to understand how that fundamental difference influences its basic concepts (its notion of audience, for instance) and its modes of production (for example, digital design practices).

"The long tail" refers to the image in figure 9.1—which Chris Anderson made famous in his *Wired* article by the same name. The long-tail chart illustrates the difference between two kinds of business models: the market of hits (the left side of the chart, the Head) versus the market of niches (the right side of the chart, the Long Tail). In the old twentieth-century economy of "the hit"

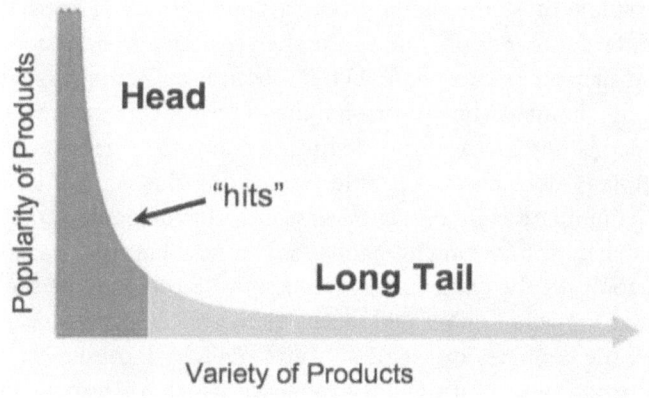

Fig. 9.1 The long tail of digital economics

(Anderson, "Rise and Fall"), businesses would develop and market a product to appeal to a large consumer base—the mass market. They would sell one kind of product, or a narrow range of products, and try to achieve high sales volume.

In the twenty-first century, Anderson argues, we are moving toward the business model of "the long tail," where "the future of business is selling less of more" (the subtitle of Anderson's book). Meaning this: We are now entering an economy where it is possible to produce a much wider variety of products and services, sell them to a small number of customers, and still succeed. There is money to be made in the long tail, where total sales can equal the total of the "head."

Some examples: According to Anderson, 98 percent of the products in a digital economy generate some sales: "a quarter of Amazon's book sales come from outside its top 100,000 titles. . . . [That is the] power of the aggregate market" (*The Long Tail* 23). Rhapsody, the online music distribution system, can compete favorably with Wal-Mart in terms of music sales, precisely because it works the economic niche of the long tail.[6] A given Wal-Mart outlet has a limited number of songs available in inventory (in hard CDs)—and so it caters to the mass market by stocking a few number of likely best-selling hits, not a wide variety of CDs. Rhapsody, as an online service, can provide more choices for consumers because it deals in digital copies and it uses the Internet as its distribution mechanism. The result is more diversity and more choices for consumers because the market can support products and services of interest to small market niches. Because of the low cost of Internet production and distribution, there is money to be made in the long tail.

We see abundant examples of this form of business on the Internet—but Anderson points out that this is not an economic development for Internet products only. His example of a non-digital product: flour. Have you noticed that there are more different brands of flour on the shelves of the grocery store now compared to ten or twenty years ago? Why? Because collecting digital information about food shoppers' buying habits allows stores to stock more precisely what is needed at any given time—in short, to cater to specific niche buyers. Digital technology does play a key role in enabling this economy to happen, as checkout scanning technology enables stores to manage inventory more precisely to match consumer buying habits. So we now have the emergence of what Anderson calls the "long-tail aggregator"—"a company or service that collects a huge variety of goods and makes them available and easy to find." Online we have the examples of Rhapsody and iTunes with music, Netflix with movies (competing successfully with Blockbuster, which has responded by developing an online component), eBay and Amazon with physical goods, and Google with information.

This approach to marketing products represents a philosophical shift away from "one size fits all" thinking and the mentality of "the big hit" to products and services tailored for smaller groups, for specific user needs, for market niches. In other words, we are talking about a different kind of consumer "audience," as both Anderson and Yochai Benkler acknowledge: "The audience is shifting to something else. . . . Increasingly the mass market is turning into a market of niches. . . . thanks to the economics of digital distribution" (Anderson, *The Long Tail* 5–6). "Consumers are changing into users—more active and productive than the consumers of the industrial information economy" (Benkler, *Wealth* 126–27).

This development creates the expectation for online writing that information will be tailored to specific user needs and to small audiences: I could do that if time and resources were not an issue. But I'm running a software company; how can I afford to write twelve different user manuals for twelve different kinds of users? How do I deliver tailored information products to that long tail of users? The answer lies, I think, in repackaging (that is, single sourcing) and outsourcing. You take existing information and redesign it, remediate it, and redeliver it for new audiences and purposes. You produce RIOs (reusable information objects). You do not develop new content yourself, or at least not very much. Rather, you "outsource" by setting up and managing a social network for users.

The software industry has been using this approach for some time with online user forums. In the old model of technical writing, you would write a comprehensive user manual that tried to imagine every task, every problem that a user might encounter using an application or product. That is, the one-size-fits-all, big-hit approach that generates a 462-page manual (that nobody reads). Then, to handle particular user questions, you would set up a telephone help line. It became clear early on that such an approach was not economically viable—it was expensive, and it did not work that well for solving specific user problems.

What we see now is a different approach. If you are a professional communicator thinking about how to provide user support, you might develop a three-pronged approach: (1) Provide minimal basic print documentation that everybody is likely to need. (2) Provide online tutorials to help people learn to do specific tasks. (3) Sponsor a user community where users can help one another answer very specific, idiosyncratic questions. This community network links users of the application or product. The network can provide conventional information resources (documentation, tutorials), but it also has a user forum that allows the users to help one another out. The professional communicator's role in this process is to design the forum, provide editorial controls, develop new documentation as needed, and add functionality to the site. This is using the long tail in conjunction with a social, user-based approach to documentation.

An example of such an approach is Adobe's user forums. Users of particular software applications, such as Dreamweaver, can join a threaded discussion related to use of the product.[7] I logged in to one of these Dreamweaver support forums at 9:30 one Sunday morning and found thirty-five other users logged in. A group of four was helping "SuzyQ2U" solve a problem she was having with installing pop-up menus. She had posed a very specific question that concerned not only Dreamweaver but also her particular network configuration, her browser, and other issues particular to her local context. The problem was a highly specific and local one, in other words. Other users helped her solve her problem. Another time, I visited the Dreamweaver forum at 7:00 A.M. I found seventeen users logged in to the General Discussion site. Two users—cripaustin and Murray—had an exchange of several postings just between the two of them: cripaustin had a "Quick Question" about centering divs, and Murray provided him an answer eighteen minutes later. Over several messages, they exchanged coding suggestions. In yet another discussion thread, a poster named malcster posted a message asking for critical feedback on his Web site. Over the next two days he received fifty-eight responses from nineteen different respondents.

Here is where we see the linkage between digital economics and social networking. That is the connection that Yochai Benkler has been exploring in his work, particularly in *The Wealth of Networks*. Benkler is investigating this phenomenon of social sharing in terms of gift exchange economy (see also Benkler, "Political Economy"; Benkler, "Sharing Nicely"). His first point is that conventional monetary notions of economics are inadequate for explaining the phenomenon of social networking. Like carpooling, social networking does not usually generate dollars directly—but, like carpooling, it does generate economic value, value that is not easily captured by standard economics models.

The term that Benkler employs to describe this phenomenon is "commons-based peer production," a term that describes a mode of economic production in which the creative energy of large numbers of people is coordinated (usually with the aid of the Internet) into meaningful projects, mostly without traditional hierarchical organization or financial compensation.[8] He compares this mode of production to "firm production" (wherein a centralized decision process decides what has to be done and by whom) and to "market-based production" (in which tagging different prices to different jobs serves as an attractor to anyone interested in doing the job).

The key social feature of such an economic model is the presence of a "commons," which Benkler defines as "a particular type of institutional arrangement for governing the use and disposition of resources. Their salient characteristic, which defines them in contradistinction to property, is that no single person has exclusive control over the use and disposition of any particular resource. Instead, resources governed by commons may be used or disposed of by anyone among some (more or less well defined) number of persons, under rules that may range from 'anything goes' to quite crisply articulated formal rules that are effectively enforced" (Benkler, *Wealth* 2).

The Dreamweaver user forums might be seen as an example of a commons, and also as an example of "distributed writing"—to adapt the notion of "distributed computing," the technique of deploying multiple processors working in tandem to solve computational problems (Benkler, "Sharing Nicely" 289). Distributed writing refers to solving human problems by creating a viable commons, a social network that will tap into the wisdom of the crowd.[9]

## Social Networking and "The Wisdom of the Crowd"

*Time* magazine's "Person of the Year for 2006" was "You"—by which *Time* meant everybody who engages in social networks on the World Wide Web and contributes value to those networks.[10] Of course "You" is not "everybody"—so the "you" here already exposes a gap in the conversation. "You" is actually a privileged minority of higher-end users. Many of the Web 2.0 advocates assume that "everybody" is involved in social networking. Clearly not so.

What *Time* is acknowledging is the power and value of social networks: a lot of people are using the Internet to share information, even when they are not being paid for it. Social networking refers to sites and applications that create a user community and that allow (more accurately, depend on) users to produce content—that is, their existence depends on UGC (user-generated content). Users upload, store, and tag content (bookmarks, videos, photos). This creates a large searchable and *dynamic* database that all users in the community can access. This kind of social network is a "folksonomy" (Joshua Porter), a database in which the community of users (including so-called nonexperts) contribute content and create the organizing structure through tagging. This type of Web design taps into "the wisdom of crowds." It is unlike a taxonomy, for which the organizing structure is predetermined, top-down, and expert driven.

The chief advantage of a folksonomic approach is that it allows you to "see what people are thinking," to find out what people are reading, and to see what tags others use to organize content. It puts the wisdom of the crowd to work for the community. The other key to it is social interaction. It is not just a static, one-way, top-down delivery of prefab information. The information is in constant flux, and there is a constant social interaction involved in the process of sharing it.[11]

The key feature of these social sites is the tags that users use to label files—a kind of metadata similar to a keyword but created by the person uploading the file (that is, ordinary users). This approach is based on the assumption that "crowds have wisdom," which is James Surowiecki's main point in *The Wisdom of Crowds*. Surowiecki's basic argument is that "under the right circumstances, groups are remarkably intelligent, and are often smarter than the smartest people in them" (xiii). "The right circumstances" is an extremely important phrase. For certain specialized skills, the crowd is not the best approach. Think about auto mechanics, brain surgery, and airplanes in flight. For those functions, you certainly want domain experts, specialists with particular knowledge and skills pertaining to that particular task.[12]

The crowd can function better than domain experts, according to Surowiecki, in larger, messy, unpredictable, and complex social problems and decision making. One example Surowiecki cites is the beehive (referencing and echoing Thomas Seeley's book *The Wisdom of the Hive*). Bees cooperatively search for food, pool their information, and maximize the resources of the community to locate the optimal food source to ensure the success of the group: "Bee foragers end up distributing themselves across different nectar sources in an almost perfect fashion, meaning that they get as much food as possible relative to the time and energy they put into searching. It is a collectively brilliant solution to the colony's food problem" (Surowiecki 27).

Another example is the Pro-Am movement in astronomy, a movement in which professionals and amateurs work together to create more reliable knowledge, and create it faster, than professionals alone (because more eyes matter): "Amateurs multiply the power of astronomy many times" (Anderson, *The Long Tail* 61).

Another example comes out of the research of one of my dissertation students at Purdue University, Laurie Cubbison, who conducted an online ethnography studying patients with chronic fatigue syndrome and fibromyalgia syndrome ("Validating Illness"). Cubbison studied an online community developed by patients for sharing information and for talking about their condition because they were not getting the help they needed from their doctors. They pooled their collective knowledge about symptoms and treatments in a way that actually created clinical knowledge about chronic fatigue and fibromyalgia. In isolated locations, neither doctors nor patients had enough knowledge or experience. But through the power of the collective the group was able to create useful medical knowledge, raising awareness about the syndromes and about which treatments worked and which did not.

So the question is, In what positive ways can the crowd contribute to the composing process? We can quickly call on some general principles: Many minds are good for brainstorming and project conception. However, too many cooks can spoil the drafting process. For some phases of written production, the crowd has a positive value; for others, it is better to deploy experts and individuals. But this question merits more composition research: When does folksonomic involvement help writing production? When is individual production more effective and efficient?

### Social Networking Case: The "Teachers for a New Era" Project

The Teachers for a New Era (TNE) project at Michigan State University provides an example of how social networking can help solve a complex social problem.[13] It shows how writers can design a social system aimed at distributing complex information to a broad user base in a way that will be useful to those users. The key to success here is twofold: (1) In accordance with assumptions of folksonomic social networking, let the users decide what is valuable information. Do not impose value from the top down. Do not let the domain experts overdetermine value. Rather, circulate information throughout the user pool and let the wisdom of the crowd determine value. (2) Create a system that helps users do their work more efficiently. Reduce the learning curve for participation. Create a participatory social economy that generates value through increased participation. Again, Web-based social networking designs are ideal for meeting these criteria.

The five-year TNE project was funded by the Carnegie Foundation with the goal of developing a comprehensive and rigorous set of teacher knowledge standards for teacher education at Michigan State University—with the ultimate goal of systemically changing the structure of teacher education nationwide. The major outcome of this project was the Teacher Knowledge Standards (TKS) guide, a comprehensive set of standards for K–12 teacher education across a number of subject areas: science, math, social studies, and literacy education. The TKS was published in November 2004 in the Green Book, a thirty-four-page bound print document presenting several hundred standards, in outline form, across six major topic areas.

I came into the project as part of a consulting team coordinated through the WIDE Research Center at Michigan State University and was charged with this task (in year 4 of the TNE project): Figure out a way to deliver the TKS to the intended audience—teacher educators in a variety of subject areas—in a way that would give the TKS persuasive power and influence. In other words, we were called on to play a fairly traditional technical communications role: serve as a

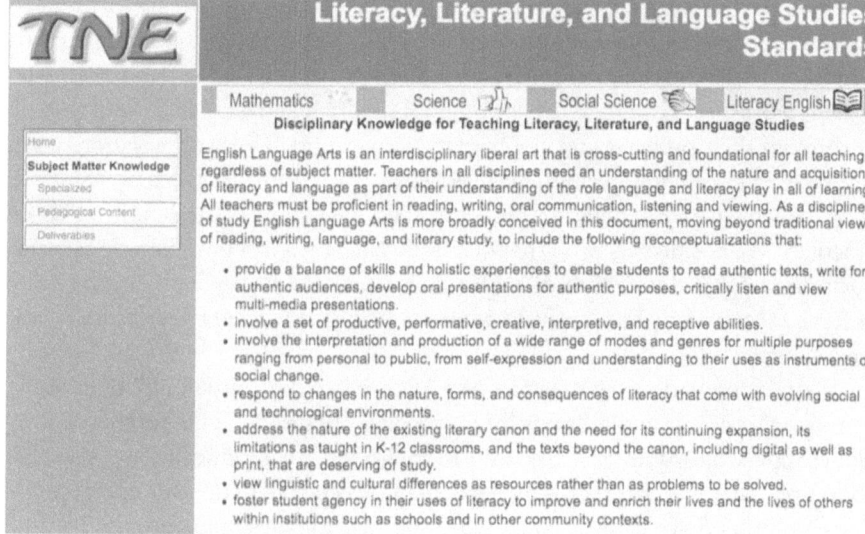

Fig. 9.2 Web 1.0 online version of Teacher Knowledge Standards for Literacy, Literature, and Language Studies

bridge or translator, transmitting expert-developed knowledge to end users expected to implement that knowledge.

The Green Book was clearly not an effective way to deliver the TKS—and we observed that early in the first phase of our research. Print standards are inert and static. They do not influence or integrate well with teachers' work. On the practical level of writing production, teachers find it difficult to move (copy and paste) standards from print format into forms that they actually use on a daily basis (assignment descriptions, teacher evaluation rubrics). In addition, there was considerable political resistance to the TKS. Most teachers' responses to the idea of the TKS were, yeah, yeah, another set of standards—No standards left behind. They already saw themselves as forced to deal with multiple standards, multiple bureaucracies producing standards, mostly by outside agents with little or no knowledge of (or respect for) the work that teachers actually do. This was not a promising communication situation.

Simply uploading the Green Book to the Web did not help either.[14] On the Web, teachers could more easily view the standards and more easily copy and paste them into instructional materials. But the standards still remained inert and abstract and dauntingly textual, as shown in figure 9.2.

Our recommendation derived from our interviews with teachers and our observations of their work practices. In other words, our research process started with deep audience analysis, particularly analysis of the teachers' work practices. Our first research question was, How do teachers currently integrate standards into their work? In our interviewing we used a contextual inquiry approach.[15]

We interviewed the teachers in their offices or classrooms, we asked them to show us documents they used in their teaching, we looked at how they organized information on their hard drives, we observed where standards were located in their workspaces (on hard drives, on bookshelves), and we asked them to walk us through their processes for doing class preparation. These observations were aimed at trying to understand teachers' work practices in situ so that standards could become enmeshed in those practices and, most important, helpful to those practices. We decided early on that delivering the TKS in a one-directional, top-down model was doomed to fail. Our contextual interviews had allowed us to observe far too many standards documents sitting on teachers' bookshelves collecting dust.

We recommended the creation of a Web-based "resource hub" that would combine taxonomic and folksonomic approaches to design. The Teacher Knowledge Standards would be presented more or less taxonomically; we saw that necessity—that was, after all, what our client directed us to do. But our observations and interviews with teachers indicated that for these standards to be used and embraced, something other than a top-down, taxonomic model was necessary. The best hope for implementation of the standards was to intertwine them with teachers' work practices and create an information product that would allow teachers the opportunity to engage, contest, and revise the standards. For the TKS to have the desired influence, the teachers needed to make the standards their own; to have the ability to revise, select, translate, critique, and prioritize them; and to be able to integrate them easily into their pedagogies (for example, have easy access to useful information that could be quickly copied and pasted into lesson plans and assignment sheets).

Hence we also integrated a folksonomic design that allowed teachers to post their own assignments and syllabuses and use the research hub as a kind of content management system for their teacher education courses, but then also link their materials to the standards. We hypothesized that this dual design model would promote adoption, and also revision, of the TKS.

The product that we ultimately produced for "delivery" of the standards—the Literacy Resource Exchange—was a social network for teachers to share materials within an environment where the standards hovered implicitly as one tab among many on the interface. Teachers could engage them or not, discuss them or not, as they wished. This environment values the wisdom of the crowd and seeks its input—rather than simply mandating adherence to predetermined standards or procedures. Figure 9.3 shows the home page interface for the first version of the Literacy Resource Exchange—showing what the teacher sees after logging on to the site. This page contains a window for Popular Tags, leading the teacher to resources that others have found useful. The site allows teachers to upload and share teaching materials. It also allows them to create discussion groups and to share teaching resources within those groups.

Fig. 9.3 Web 2.0 social networking site for teacher educators—early test version of the Literacy Resource Exchange (Porter and Hart-Davidson)

The Literacy Resource Exchange is, in effect, a social network along the lines of Delicious, YouTube, or Flickr. It is based on the folksonomic assumptions that teachers have wisdom, that the best way to produce an economically valuable resource for teachers is to give them responsibility for developing its content, and that the best way to "present" the teacher standards is to put them in a dynamic system that allows them to be changed, revised, prioritized, ignored, and, in effect, re-created. As a model for the writing process, our work involved not generating content so much as (1) understanding our audience's needs, talents, and knowledge and (2) designing a "commons," a social network encouraging teachers to share their knowledge in ways beneficial to the group.

### The Dark Side of Social Production: Issues of Access and Labor

Before we embrace too wholeheartedly the benefits of social networking, peer production, and user-generated content, it is important to consider several critical-ethical questions: Who is excluded from this digital economy? Who does not have full (or any) access to participation? And who, then, is left behind? Who is rewarded and paid for their labor? And who is not? Is an economy based on so much "free labor" a fair and just economy?

### The Issue of Access

Designing for access has long been an important consideration for Web designers—certainly for ethical reasons of equity and fairness, but, increasingly, for economic reasons as well. The basis of long-tail economics is creating content for small market niches—and doing that requires designing information for particular user needs. It has never been ethically fair—and it is now no longer economically smart—to design systems for some standard "generic user."

Access is an old problem in rhetoric. We could see it as the problem of addressing the general audience, a question that the eighteenth-century rhetorician George Campbell addressed in *The Philosophy of Rhetoric* (see James E. Porter, *Audience* 32–34). It is easy to look at general-population characteristics and make facile judgments about "access." In practice, though, access is a complex challenge that rhetoric ought to be taking up more earnestly, as it pertains in basic ways to audience. Accessibility is an important consideration for system design—and not simply accessibility for physically challenged users. What the data show is that there are a range of accessibility variables that need to be accounted for, ranging from socioeconomic status, to broadband access, to sight issues, to educational limitations.[16]

It is not enough to say that Internet usage is now "widespread," just because we have data that tells us there are 200 million Internet users in the United States and that 75 percent of adults in the United States use the Internet "at least occasionally" (Pew). Those grand numbers by themselves mislead. Within these generalized numbers lie some troubling educational and socioeconomic differences. It is important to note that, as of August 2008, for those with household incomes less than $30,000 per year the level of Internet usage is only 56 percent, whereas it is 95 percent for those with annual household incomes greater than $75,000. And although the overall percentage of U.S. residents using the Internet is rising, the gap between users and nonusers is widening. Only 38 percent of Americans over age 65 use the Internet. And, maybe most troubling of all, only 38 percent of those who have not graduated from high school use the Internet (the figure is 66 percent for high school graduates and 95 percent for college graduates; Pew; Fox). We can perhaps expect that the age divide will lessen over time, but the education and socioeconomic divides seem to be widening, not narrowing.

Designing for a diverse audience is a challenge that needs to be addressed—and even principles of universal design may not be sufficient to address this wide variety of needs. The first design principle is an equity issue: All interested users should be able to participate in social networks. The second principle is an economic one as well as an ethical one: Diversity is a key criterion for design of systems that rely on the wisdom of the crowd; "the simple fact of making a group diverse makes it better at problem solving" (Surowiecki 30).

## The Issue of Labor

Richard Barbrook ("Digital Economy"; "High Tech") and Tiziana Terranova are two cultural critics, among many, who urge caution regarding the optimistic claims about open source and social networking bringing new power to users: "We cannot agree with the digerati's claims that the Internet turns every user into an active producer, and every worker into a creative subject" (Terranova). They wonder whether a generation of Netslaves, a new form of oppressed labor, is being duped into volunteering their expertise. Terranova wonders whether the increasing reliance on user-generated content results in "a degradation of knowledge work" and a devaluing of digital skills.[17]

When Doritos invited consumers to make their own video commercials[18] (as part of its "Crash the Superbowl" marketing campaign), and then announced its plan to air the winning commercial during the 2007 Super Bowl XVI broadcast, was that an innovative approach to advertising that provides an opportunity for the ordinary user? Or was it a form of outsourcing that takes jobs away from knowledge workers? You have to be a little worried when *BusinessWeek* pronounces "free labor" as the Best Idea of 2006 and goes on to suggest that business should take full advantage of this windfall before the digital dupes wake up: "How long before the unpaids start stomping for their cut? Catch them while you can" ("Free Labor").

Not all gift economies are innately unjust, as Terranova admits: "Free labor is not . . . necessarily exploited labor." Think about volunteer fire departments, academic discussion groups, or unpaid student internships. But we need an ethical metric for determining when labor is exploitative versus when labor works to mutual benefit, to generate value for all parties. One such metric is a reciprocity principle: What value do workers derive from free labor? Lerner and Tirole point to two incentives motivating programmers to contribute free labor in the open source movement, the career concern incentive and the ego gratification incentive: "The *career concern incentive* refers to future job offers, shares in commercial open source-based companies, or future access to the venture capital market. The *ego gratification incentive* stems from a desire for peer recognition" (58). Lerner and Tirole also note that programmers can gain practical knowledge that can directly translate into cost savings, but that the greater long-term benefit might be increased systems knowledge from working in a fluid environment with a larger number of other contributing programmers.

Barbrook ("Digital Economy") points out that a new kind of worker—the "digital artisan"—emerges from this kind of economy, and that such a person develops skills that can translate into pay eventually. In Anderson's terms, the digital artisan may start out at the far end of the long tail, but through experience and exposure and circulation, become more and more well known, move toward the body, and achieve financial success. Although this narrative sounds promising, we should be wary of such Horatio Alger stories. The process might work for a few, but does it work for most?

The metric might be qualified, then, to something more like "immediate and comparable reciprocity"—that is, the value is obtained fairly soon and at a comparable level of exchange. Think about the gift economy of the academic discussion list. Academics within such lists typically post questions to the group (for example, asking about resources on a given topic), and benefit from the wisdom of the crowd in the form of helpful information collected from an expert community. But what are the ethical expectations governing such communities? If you read the list and benefit from the wisdom of the crowd, are you expected to return the favor, at least occasionally, when somebody else in the community needs information that you could provide? If the flow of information is only one way, a few committed participants providing free information, then is the community at large exploiting the good will and commitment of the few? Or do the members of such communities post out of sheer good will and the satisfaction of helping others? Is reciprocity not necessarily an expected part of such an economy? Is the value to be gained in the giving rather than in the receiving?

This kind of gift-sharing economy does not generate revenue internally: nobody gets paid for posting. But the internal economy does generate revenue externally—if you think in terms of increased knowledge and productivity. A forum can lead to improved professional status for its participants in the form of publications, jobs, promotions, consulting work, teaching tips, practical skills gained, and the like. Are user forums sponsored by companies such as Adobe working on a similar ethic—or are those user forums an instance of the software industry cutting costs, outsourcing labor, and avoiding its responsibility to support its products? Are the companies simply offloading technical help on to their customers? Or do such forums supplement more traditional forms of support by deploying the wisdom of the crowd to solve problems and answer questions that conventional documentation and online help could not as efficiently address?[19]

Those who design interactive systems must ask such critical-ethical questions—and, beyond merely asking them, follow up to make sure that social networks meet relevant accessibility standards and, if the systems rely on user-generated content, that they provide reciprocal value to users, not simply take advantage of free labor.

## Conclusion: Implications for Writers

When talking about writing in digital spaces, we need to reconceptualize writing from the economic standpoint of production, consumption, and exchange (Trimbur; Marx). Writers in the digital milieu encounter an economic exchange system that is different from that of print. Capital resides not so much in the original texts you produce, but rather in your ability to deliver and circulate texts in ways that make them accessible and useful to others and in your ability to collaborate with others, to share files, to co-create meaning in social spaces. In other words, in the digital economy, what we come to think of as "writing ability" is shifting in rather dramatic ways toward a community and collaborative notion of networked writing. The professional writer becomes more a creator of communities, of networks, than a creator of content.

In the field of technical communication Johndan Johnson-Eilola has been talking about this shift for ten years or more, noticing, first, that the emphasis is shifting from "technical" to "communication": that is, digital industries have "shifted portions of their revenue streams to providing information rather than technological products. Some organizations that work specifically in information produce little or no products of the industrial type" (Johnson-Eilola, *Datacloud* 252). In this information economy the professional writer becomes what Johnson-Eilola calls, citing Robert Reich's term, a symbolic-analytic worker: "People in this type of work identify, rearrange, circulate, abstract, and broker information in response to specific, concrete situations. They work with information and symbols to produce reports, plans, and proposals. They also tend to work online, either communicating with peers . . . or manipulating symbols. . . . Creativity is no longer the production of original texts, but the ability to gather, filter, rearrange, and construct new texts—symbolic-analytic work, articulations" (Johnson-Eilola, *Datacloud* 28, 134).

The traditional assumption that expertise lies mainly within disciplinary or professional domains of expertise is challenged by Web 2.0 developments, advocates of which advance the counter position that expertise lies in community choice through "folksonomic tagging" (Shirky, "Communities"; Shirky, "Power Laws"; Joshua Porter; O'Reilly, "What Is Web 2.0?"). In the world of Web 2.0, information content development is determined by communities of users, ordinary end users who are not experts or domain specialists (see Brown and Duguid; Golder and Huberman). For certain kinds of tasks—particularly for messy, complicated, and open-ended exploratory work and for solving wicked problems—the best option may be to get experts out of the way and to design systems to include dynamic collaboration with nonexperts.

The field of technical communication has long lived with the distinction between domain expert (aka, content producer, specialist, scientist) and end user (aka, audience, public, nonspecialist). This dichotomy between expert/producer and nonexpert/receiver is a key defining feature of the field, historically speaking (for example, it is the fundamental assumption of the linear communication model, the so-called Shannon-Weaver model[20]). Not coincidentally, it has also been the defining historical binary for the field of rhetoric since the fifth century B.C.E.—that is, the distinction between rhetor/writer and audience. In technical communication the process of user-centered design is an acknowledgment that users have useful knowledge and can contribute productively to the design of systems. User-centered design is a development model geared toward bringing user knowledge into the design process at a much earlier stage (Johnson).

The social dynamic of Web 2.0 threatens to overturn the fundamental expert-novice rhetorical model upon which writing and communication theory has long depended. What if the job of experts is not to solve problems by themselves, but rather to design robust collaborative systems that allow diverse groups of users (experts and nonexperts alike) to pool community resources in order to solve problems? The notion of "expert" is shifting away from its traditional basis in "content knowledge" to another basis: expert as skilled social networker and collaborator.

In the kind of digital economy discussed here the role of the writer shifts from production of original content for a large mass audience to managing information resources in ways that direct tailored information to smaller audiences, maybe even the single user, with very specific needs. Doing this work requires knowledge of how to do research (particularly audience research), how to work within and to manage collaborative teams, how to deliver information, and how to test and evaluate information (usability knowledge).[21]

As if the author was not already dead, Web 2.0 fires more bullets into the author's cold carcass. The rhetorical shift occasioned by Web 2.0 creates a technological presumption in favor of end users (audiences). In such a writing economy, some traditional writing skills continue to be important—research, audience analysis, rhetorical effectiveness, collaboration. But these practices work in very specific ways in online environments—and, I would argue, cannot be effectively taught outside those environments. Overall, the skill set that is needed for work in this economy is the ability to

- repackage, redesign, remediate, and redistribute existing information for new audiences and contexts;
- make and maintain connections (a) between people, and (b) between people and information resources;
- design social networks that enable productive collaborative thinking and work and that allow for the effective and efficient distribution of information;
- select and tailor information for small market niches (specific audiences); and
- design indexing, tagging, filtering, and searching strategies that allow audiences to find needed information.

Finally, a question: Is the appropriate role for rhetoric simply to follow technology development—to adapt its theories and practices to fit changing communication circumstances? Or is it possible that rhetoric can help shape and influence the digital economy and social networking? My answer to that question can be summed up in two phrases: "information" and "knowledge work." If the basis of a digital economy concerns (a) the development of "information"—and not just information as a static product, but more important the transformation of information into useful knowledge; and (b) if the digital economy concerns the delivery and circulation of information via social networks in ways that create value for users, then writing teachers, communication scholars, and rhetoric theorists certainly have a lot to offer this discussion. The neglected rhetorical canon of delivery again becomes important. Not the old version of delivery for oral discourse, but a remediated delivery for digital environments. To accomplish this, rhetoric must understand why and how digital economics pertain to writing practice and shift its theoretical and pedagogical emphasis toward digital forms of invention, production, and interaction. In 1990 Kathleen Welch admonished us about a new paradigm we still have not quite heeded: "The fifth canon [delivery] . . . is now the most powerful canon of the five" (31).

## Notes

1. According to Cormode and Krishnamurthy, "the essential difference between Web 1.0 and Web 2.0 is that content creators were few in Web 1.0 with the vast majority of users simply acting as consumers of content, whereas any participant can be a content creator in Web 2.0 and numerous technological aids have been created to maximize the potential for content creation." See also O'Reilly, "What Is Web 2.0?"; O'Reilly, "Open Source"; and the Wikipedia definition of Web 2.0 at http://en.wikipedia.org/wiki/Web_2.

2. Other important discussions of digital economics include Barlow; Raymond; Lessig; Tapscott; and Williams, "Innovation"; Tapscott and Williams, *Wikinomics*. See also "The Power of Us."

3. See James E. Porter, "Why Technology," "Rhetoric," "Why We," "Opening," "Recovering"; Porter and DeVoss, "Rethinking"; and DeVoss and Porter, "Why Napster."

4. Writing well before the digital age, Pierre Bourdieu tells us two things of importance to digital distribution: (1) the importance of symbolic capital in a society should never be underestimated; and (2) the relationship between symbolic and material capital matters (that is, they have an effect on each other). Symbolic capital is tied to the potential and actual development of material economic capital.

5. Nardi et al. ("Why We Blog") conducted in-depth interviews with twenty-three bloggers to determine their motivations. What they found was a variety of motivations for blogging, including "documenting one's life; providing commentary and opinions; expressing deeply felt emotions; articulating ideas through writing; and forming and maintaining community forums" (43). The value of blogging, for most of these bloggers, pertained to their desire to articulate and share their views; monetary gain was not a principal motivating factor.

6. For a visual graphic of the long tail as it applies to Rhapsody online music sales, see http://longtail.typepad.com/the_long_tail/images/tailgrowth2_1.jpg.

7. For examples of such user forums, see "Dreamweaver Support Forums" at http://www.adobe.com/cfusion/webforums/forum/index.cfm.

8. Most academics are involved in commons-based peer production, or at least they are if they participate in email discussion groups (aka, listservs). Most professional discussion groups—such as H-RHETOR (for scholars working in the history of rhetoric), AoIR-L (for the Association of Internet Researchers), and CHI-WEB (for Web designers)—are based on a gift-exchange economy. Scholars and practitioners participate on these lists not to make money directly but rather to share information and resources of value to the community. You post information helpful to others in the hope (or expectation) that you will receive useful information in return. But sharing is not

the only motive in this kind of context: You might also hope to establish your reputation, to become known and respected as knowledgeable in a certain area, to distribute and circulate your own work, or to enhance your scholarly capital. You join lists pertinent to your interests, your research, your teaching, your political aims—and you contribute according to interest and value. No money ever passes hands on these lists. But such lists are common and active and, I would argue, useful.

9. Of course in any social network based on a gift-sharing economy, there are always "freeriders" (Ripeanu et al.), lurkers, and low-sharing users—participants who tap into the knowledge of a community without contributing any value themselves. As Ripeanu et al. discovered, "Our data for the distribution of contributions within a single community shows that a minority of gifters in a community are responsible for most of the gifting." However, they add, digital communities can install social protocols to encourage gifting. Nielsen also points out that within most user groups "participation inequality" is rampant. He refers to the 90–9–1 principle: in most online forums 90% of users lurk, 9% contribute occasionally, 1% produce the bulk of the information. On Wikipedia, 99.8% of users are lurkers.

10. To view the *Time* magazine cover for December 26, 2005, visit http://www.time.com/time/covers/0,16641,20061225,00.html.

11. Some examples of folksonomic Web 2.0 Web sites include Delicious (a site for sharing bookmarks), Flickr (for photo management and filesharing), SlideShare (for slide presentations), and YouTube (for videos).

12. Surowiecki is careful to delineate circumstances in which the wisdom of the crowd can effectively be deployed, versus circumstances in which "the madness of the mob" is likely to prevail. The wisdom of the crowd works well for complex social problems of an interdisciplinary nature. But to deploy this wisdom appropriately requires designing a social network based on these features: diversity of opinion (skills, knowledge); independence; decentralization; aggregation; access to information; and simultaneous (not sequential) decision making.

13. The TNE project Web site is at http://tne.msu.edu/default.htm.

14. For a complete list of the Teacher Knowledge Standards, see https://www.msu.edu/~tne/index.html.

15. Contextual inquiry is a user-centered methodology that involves interviewing and/or observing users in their normal workspaces and, to the extent possible, observing their work practices.

16. For a more detailed discussion of access as a rhetorical subtopic of delivery, see James E. Porter, "Recovering Delivery."

17. Søren Mørk Petersen worries about "capitalism's ability to piggyback" on user-generated content. He outlines the ways in which Web 2.0 can be viewed as "an architecture of exploitation that capitalism can benefit from: 1. Through a distributed architecture of participation, companies can piggyback on user generated content by archiving it and making interfaces, or using other strategies such as Google's AdSense program. 2. Designing platforms for user generated content, such as Youtube, Flickr, Myspace and Facebook." See also Albrechtslund; Scholz.

18. See http://promotions.yahoo.com/doritos/index.php.

19. For example, for its popular Web-authoring tool Dreamweaver, Adobe provides a fairly robust Help Resource Center in addition to its user forums. In other words, the

company is still producing documentation and tutorials (some of which permit user comments and annotations) in addition to sponsoring user-generated assistance. In this case, I would say, Adobe is not abdicating its responsibility to users but rather deploying the wisdom of users, in conjunction with conventional modes of help, to provide the best possible range of assistance.

20. For a visual representation of "The Shannon-Weaver Model" of communication see http://www.cultsock.ndirect.co.uk/MUHome/cshtml/introductory/sw.html.

21. Research and theory pointing us in the right direction includes Bonnie Nardi's research on network WORK (see Nardi, Whittaker, and Schwarz, "It's Not" and "NetWORKers"); Johndan Johnson-Eilola's research on the changing nature of work in technical communication (Johnson-Eilola, "Accumulation," "Relocating," and "Writing"); and William Hart-Davidson's emergent work ("Web 2.0"). Hart-Davidson ("Web 2.0") asks the important question, What happens to writers when users become content producers? (See also Hart-Davidson, "On Writing.") Others in writing studies who have discussed the implications for writers of Internet-based technology development and new media include Daniel Anderson; Rice; Hoffman; Reid; and WIDE.

## Works Cited

Aigrain, Philippe. "The Individual and the Collective in Open Information Communities." *Free/Open Source Research Community*. 2003. Accessed 10 July 2008, http://opensource.mit.edu/papers/aigrain3.pdf.

Albrechtslund, Anders. "Online Social Networking as Participatory Surveillance." *First Monday* 13.3 (2008). Accessed 7 December 2008, http://www.uic.edu/htbin/cgiwrap/bin/ojs/index.php/fm/article/viewArticle/2142/1949.

Anderson, Chris. "The Long Tail." *Wired* 12.10 (2004). Accessed 10 July 2008, http://www.wired.com/wired/archive/12.10/tail.html.

———. *The Long Tail: Why the Future of Business Is Selling Less of More*. New York: Hyperion, 2006.

———. "The Rise and Fall of the Hit." *Wired* 14.07 (2006). Accessed 10 July 2008, http://www.wired.com/wired/archive/14.07/longtail_pr.html.

Anderson, Daniel. "Web 2.0 and Writing Outcomes." *I Am Dan*. Posted 28 July 2006, accessed 10 July 2008, http://sites.unc.edu/daniel/2006/07/web_20_and_writing_outcomes.html.

Barbrook, Richard. "The Digital Economy: Commodities or Gifts?" *Nettime*. Posted 17 June 1997, accessed 10 July 2008, http://subsol.c3.hu/subsol_2/contributors3/barbrooktext.html.

———. "The High-Tech Gift Economy." *First Monday* 3.12 (1999). Accessed 10 July 2008, http://www.firstmonday.org/issues/issue3_12/barbrook/.

Barlow, John Perry. "The Economy of Ideas." *Wired* 2.03 (1994). Accessed 10 July 2008, http://www.wired.com/wired/archive/2.03/economy.ideas.html.

Benkler, Yochai. "The Political Economy of Commons." *Upgrade* 4.3 (2003): 6–9.

———. "'Sharing Nicely': On Shareable Goods and the Emergence of Sharing as a Modality of Economic Production." *Yale Law Journal* 114 (2004): 273–358.

———. *The Wealth of Networks: How Social Production Transforms Markets and Freedom*. New Haven, Conn.: Yale University Press, 2006.

Bourdieu, Pierre. "The Forms of Capital." Translated by Richard Nice. In *Ökonomisches Kapital, Kulturelles Kapital, Soziales Kapitaln. Soziale Ungleichheiten*, edited by Reinhard Kreckel, 183–98. Goettingen, Ger.: Otto Schartz, 1983. Accessed 7 December 2008, http://www.knowledgepolicy.com/2005/08/bourdieu-forms-of-capital.html.

Brown, John Seely, and Paul Duguid. "The Social Life of Documents." *First Monday* 1.1 (1996). Accessed 10 July 2008, http://www.firstmonday.dk/issues/issue1/documents/.

"The Power of Us: Mass Collaboration on the Internet Is Shaking Up Business." *BusinessWeek*, 20 June 2005. Accessed 8 July 2008, http://www.businessweek.com/magazine/content/05_25/b3938601.htm.

Carter, Locke. "Rhetoric, Markets, and Value Creation: An Introduction and Argument for a Productive Rhetoric." In *Market Matters: Applied Rhetoric Studies and Free Market Competition*, edited by Locke Carter, 1–52. Cresskill, N.J.: Hampton, 2005.

Cormode, Graham, and Balachander Krishnamurthy. "Key Differences between Web 1.0 and Web 2.0." *First Monday* 13.6 (2008). Accessed 10 July 2008, http://www.uic.edu/htbin/cgiwrap/bin/ojs/index.php/fm/article/viewArticle/2125/1972.

Cubbison, Laurie B. "Validating Illness: Internet Activism in Response to Institutional Discourse." Ph.D. diss., Purdue University, 2000.

DeVoss, Dànielle Nicole, and James E. Porter. "Why Napster Matters to Writing: File-sharing as a New Ethic of Digital Delivery." *Computers & Composition* 23 (2006): 178–210.

Eyman, Douglas. "Digital Rhetoric: Ecologies and Economies of Digital Circulation." Ph.D. diss., Michigan State University, 2007.

Fox, Susannah. "Digital Divisions: There Are Clear Differences among Those with Broadband Connections, Dial-up Connections, and No Connections at All to the Internet." Pew Internet & American Life Project. October 2005. Accessed 10 July 2008, http://www.pewinternet.org/PPF/r/165/report_display.asp.

"Free Labor: Best Idea." *BusinessWeek*. Accessed 24 January 2009, http://images.businessweek.com/ss/06/12/1207_bestideas/source/2.htm.

Golder, Scott A., and Bernardo A. Huberman. "Usage Patterns of Collaborative Tagging Systems." *Journal of Information Science* 32.2 (2006): 198–208. Accessed 10 July 2008, http://jis.sagepub.com/content/vol32/issue2/.

Goldhaber, Michael H. "How (Not) to Study the Attention Economy: A Review of *The Economics of Attention: Style and Substance in the Age of Information*." *First Monday* 11.11 (2006). Accessed 10 July 2008, http://firstmonday.org/issues/issue11_11/goldhaber/index.html.

Hart-Davidson, William. "On Writing, Technical Communication, and Information Technology: The Core Competencies of Technical Communication." *Technical Communication* 48.2 (2001): 145–55.

———. "Web 2.0: What Technical Communicators Should Know as Users Become Content Producers." Presentation. Southeast Michigan STC Chapter. Ann Arbor. 28 September 2006.

Hoffman, Keith M. "Writing and Web 2.0." *Intercom* 54.1 (2007): 4–7.

Johnson, Robert R. *User-Centered Technology: A Rhetorical Theory for Computers and Other Mundane Artifacts*. Albany, N.Y.: SUNY Press, 1998.

Johnson-Eilola, Johndan. "Accumulation, Circulation, Association: Economies of Information in Online Spaces." *IEEE Transactions on Professional Communication* 38 (1995): 228–38.

———. *Datacloud: Toward a New Theory of Online Work*. Cresskill, N.J.: Hampton, 2005.

———. "Relocating the Value of Work: Technical Communication in a Post-Industrial Age." *Technical Communication Quarterly* 5 (1995): 245–70.

———. "Writing at the End of Text: Rethinking Production in Technical Communication." *Proceedings of the CPTSC*. 2000. Accessed 10 July 2008, http://www.cptsc.org/proceedings/2000/conference2000.html.

Lanham, Richard A. *The Economics of Attention: Style and Substance in the Age of Information*. Chicago: University of Chicago Press, 2006.

Lerner, Josh, and Jean Tirole. "Economic Perspectives on Open Source." In *Perspectives on Free and Open Source Software*, edited by Joseph Feller et al., 47–48. Cambridge: MIT Press, 2007.

Lessig, Lawrence. "On the Economies of Culture." *Lessig Blog*. 2006. Accessed 10 July 2008, http://www.lessig.org/blog/archives/003550.shtml.

Marx, Karl. "Production, Consumption, Distribution, Exchange (Circulation)." In *A Contribution to the Critique of Political Economy*, edited by Maurice Dobb, 188–217. New York: International, 1970.

McCloskey, Deirdre N. *The Rhetoric of Economics*. 2nd ed. Madison: University of Wisconsin Press, 1998.

Nardi, Bonnie A., Diane J. Schiano, Michelle Gumbrecht, and Luke Swartz. "Why We Blog." *Communications of the ACM* 47.12 (2004): 41–46.

Nardi, Bonnie A., Steve Whittaker, and Heinrich Schwarz. "It's Not What You Know, It's Who You Know: Work in the Information Age." *First Monday* 5.5 (2000). Accessed 10 July 2008, http://www.firstmonday.org/issues/issue5_5/nardi/.

———. "NetWORKers and Their Activity in Intensional Networks." *Journal of Computer-Supported Cooperative Work* 11 (2002): 205–42.

Nielsen, Jakob. "Participation Inequality: Encouraging More Users to Contribute." *Alertbox*. Posted 9 October 2006, accessed 10 July 2008, http://www.useit.com/alertbox/participation_inequality.html.

O'Reilly, Tim. "Open Source Paradigm Shift." *O'ReillyNet*. June 2004. Accessed 10 July 2008, http://tim.oreilly.com/articles/paradigmshift_0504.html.

———. "What Is Web 2.0? Design Patterns and Business Models for the Next Generation of Software." *O'ReillyNet*. 2005. Accessed 10 July 2008, http://www.oreillynet.com/pub/a/oreilly/tim/news/2005/09/30/what-is-web-20.html.

Petersen, Søren Mørk. "Loser Generated Content: From Participation to Exploitation." *First Monday* 13.3 (2008). Accessed 10 July 2008, http://www.uic.edu/htbin/cgiwrap/bin/ojs/index.php/fm/article/viewArticle/2141/1948.

Pew Internet & American Life Project. *Demographics of Internet Users*. Posted August 2008, accessed 23 December 2008, http://www.pewinternet.org/trends/User_Demo_10%2020%2008.htm.

Porter, James E. *Audience and Rhetoric: An Archaeological Composition of the Discourse Community*. Englewood Cliffs, N.J.: Prentice Hall, 1992.

———. "Opening Remarks: What Does Writing Practice Have to Do with Economic Development?" Conference on Writing: Digital Knowledge, WIDE Research Center, Michigan State University, East Lansing, 2006. Accessed 10 July 2008, http://www.wide.msu.edu/widepapers/read.

———. "Recovering Delivery for Digital Rhetoric and Human-Computer Interaction." Unpublished manuscript.

———. "The Rhetoric of Digital Delivery: Access, Interaction, Economics." Paper presented at the Conference on College Composition and Communication, San Francisco, 2005.

———. "Why Technology Matters to Writing: A CyberWriter's Tale." *Computers & Composition* 20 (2004): 375–94.

———. "Why We Should Recover Delivery for Digital Rhetoric: Economics." Paper presented at the Computers and Writing Conference, Stanford University, Palo Alto, Calif., 2005.

Porter, James E., and Dànielle Nicole DeVoss. "Rethinking Plagiarism in the Digital Age: Remixing as a Means for Economic Development?" Paper presented at the Conference on Writing: Digital Knowledge, WIDE Research Center, Michigan State University, East Lansing. 2006. Accessed 10 July 2008, http://www.wide.msu.edu/widepapers/read.

Porter, James E., and William Hart-Davidson. "Teachers for a New Era Project, Stage 3—Final Report." WIDE Research Center. Michigan State University, East Lansing, 29 August 2007.

Porter, Joshua. "Folksonomies: A User-Driven Approach to Organizing Content." *User Interface Engineering*. Posted 26 April 2005, accessed 10 July 2008, http://www.uie.com/articles/folksonomies/.

Raymond, Eric S. "The Cathedral and the Bazaar." *First Monday* 3.3 (1998). Accessed 10 July 2008, http://www.firstmonday.org/issues/issue3_3/raymond/.

Reid, Alex. "Tuning In: Infusing Media Networks into Professional Writing Curriculum." *Kairos* 12.2 (2008). Accessed 10 July 2008, http://kairos.technorhetoric.net/redesign/12.2/praxis/reid/index.html.

Rice, Jeff. "Networks and New Media." *College English* 69 (2006): 127–33.

Ripeanu, Matei, Miranda Mowbray, Nazareno Andrade, and Aliandro Lima. "Gifting Technologies: A Bit Torrent Case Study." *First Monday* 11.11 (2006). Accessed 10 July 2008, http://www.firstmonday.org/ISSUES/issue11_11/ripeanu/.

Salvo, Mike. "There is No Salvation: Rhetoricians Working in an Age of Information." In *Market Matters: Applied Rhetoric Studies and Free Market Competition*, edited by Locke Carter, 109–33. Cresskill, N.J.: Hampton, 2005.

Scholz, Trebor. "Market Ideology and the Myths of Web 2.0." *First Monday* 13.3 (2008). Accessed 7 December 2008, http://www.uic.edu/htbin/cgiwrap/bin/ojs/index.php/fm/article/viewArticle/2138/1945.

Shirky, Clay. "Communities, Audiences, and Scale." *Clay Shirky's Writings*. Posted 6 April 2002, accessed 10 July 2008, http://shirky.com/writings/community_scale.html.

———. "Power Laws, Web Logs, and Inequality." *Clay Shirky's Writings*. Posted 8 February 2003, accessed 10 July 2008, http://www.shirky.com/writings/powerlaw_weblog.html.

Surowiecki, James. *The Wisdom of Crowds: Why the Many Are Smarter Than the Few and How Collective Wisdom Shapes Business, Economies, Societies, and Nations.* New York: Little, Brown, 2004.

Tapscott, Don, and Anthony D. Williams. "Innovation in the Age of Mass Collaboration." *BusinessWeek*, 1 February 2007. Accessed 10 July 2008, http://www.businessweek.com/innovate/content/feb2007/id20070201_774736.htm.

———. *Wikinomics: How Mass Collaboration Changes Everything.* New York: Penguin, 2006.

"Teacher Knowledge Standards for Literacy, Literature, and Language Studies." Accessed 12 February 2009, http://www.msu.edu/~tne.

Terranova, Tiziana. "Free Labor: Producing Culture for the Digital Economy." *Electronic Book Review*. 2003. Accessed 10 July 2008, http://www.electronicbookreview.com/thread/technocapitalism/voluntary.

Trimbur, John. "Composition and the Circulation of Writing." *College Composition and Communication* 52 (2000): 188–219.

Welch, Kathleen E. "Electrifying Classical Rhetoric: Ancient Media, Modern Technology, and Contemporary Composition." *Journal of Advanced Composition* 10 (1990): 22–38.

WIDE Research Center Collective [DeVoss, Dànielle, Ellen Cushman, Bill Hart-Davidson, Jeff Grabill, and James E. Porter]. "Why Teach Digital Writing?" *Kairos* 10.1 (2005). Accessed 10 July 2008, http://English.ttu.edu/kairos/10.1/binder2.html?coverweb/wide/index.html.

# Literate Acts in Convergence Culture

## Lost as Transmedia Narrative

Debra Journet

This essay examines the television show *Lost* as a product of what Henry Jenkins calls convergence culture, a phenomenon characterized not only by media convergence (flow of media over multiple platforms and the resulting migratory behavior of media audiences), but also by participatory culture (interaction between consumers and producers) and collective intelligence (collaborative pooling of resources and skills; *Convergence*). Convergence culture, Jenkins argues, creates new forms of "transmedia storytelling," in which the narrative experience is so large it moves over multiple media, and so complex readers/consumers must work together in "knowledge communities" to understand its full detail and coherence.[1]

In writing this essay, I drew on my own experience watching *Lost* and on responses to *Lost* posted on fan sites to examine some of the literate practices this new form of transmedia storytelling encourages. These practices encompass the analytic and interpretive skills that readers traditionally bring to a complex narrative text. They also entail new kinds of critical interactions among readers and authors and therefore new ways to construct and respond to narrative. Moreover, in their responses, many fans exhibit an intensely detailed and even passionate relationship to *Lost*, committing enormous amounts of time and attention to understanding, appreciating, and perhaps even shaping the fictional story *Lost* offers. All this suggests new ways to read, view, and write about narrative in convergence culture.

As of December 2008, *Lost* had completed its fourth season. Both a critical and popular success, *Lost*, at its height, attracted more than fifteen million viewers. The story line of *Lost* concerns the survivors of Oceanic Flight 815, en route from Sydney to Los Angeles, after it crashes on a mysterious tropical island. At one level, the show narrates the experiences of these survivors over their first three months on the island. At another level, flashbacks hint at who the survivors were and what they did before boarding Oceanic 815. Beginning in season 4, flash-forwards provide cryptic clues about what happens to some of the survivors after they are rescued. As the series unfolds, we also learn more about the island and its history (what the producers call its mythology)—a world made up of such inexplicable apparitions and phenomena as a black "smoke monster," a strangely reappearing set of numerals (4 8 15 16 23 42), a series of puzzling hatches left over from a failed social science research project of the 1970s called the Dharma Initiative, and an inscrutable group of Others who are discovered to be living on the island. Similarly, as we learn more about the background of the survivors, we discover connections between them and perhaps between their past lives and what happens to them on the island.

But *Lost* is also a complex multimodal and multimedia "text" (Kress) for viewers who follow its story over multiple platforms, including not only television but also the Internet and other media. Engaging with transmedia *Lost* requires active response by viewers/readers who hunt down clues and share textual, audio, and visual information on such sites as fan forums, blogs, and wikis. These new migratory demands create challenges for participants that are both logistical and interpretive. Viewers must engage in such activities as discriminating between counterfeit and legitimate information, identifying a *Lost* "author" and discerning authorial intent, identifying allusions, summarizing episodes, sampling and downloading video and audio segments, rearranging the show's complex story into an accurate chronology, speculating on cultural parallels, and debating meaning. This careful viewing is enhanced by digital technologies that allow viewers to go through bits of story, often frame by frame, zeroing in on particular details that are not apparent to casual viewers. Interpretations are then built collaboratively, as participants pool knowledge and test theories in media-rich environments.

This essay looks at certain literate practices involved in the reception and interpretation of *Lost*. In the first half, I discuss how media convergence shapes acts of reading and viewing. My focus is on the way *Lost* exploits traditional and innovative forms of narrative complexity; in particular, how it draws on the Internet and on gaming as new ways to construct and interact with transmedia stories. Later in the essay, I consider scenes of participatory culture and collective knowledge where viewers engage in processes of "writing back" to *Lost*. I concentrate on critical and interpretive acts I observed in an online community in which participants work together to understand the structure and meaning of *Lost* as a complex narrative text. My goal is to explore both the media convergence that *Lost* makes use of and the cultural patterns of "meaning making" that such convergence promotes.

## Reading/Viewing *Lost:* Transmedia Storytelling and Media Convergence

> I think that there is a very strong likelihood that, if not now, at some time in the near future we are going to find that at least some of the maze material is intentionally released on authorized sites to become part of the whole package of what constitutes the LOST experience. I think it is part of the way that the creative minds behind it all want their work to be experienced and enjoyed. . . . I was initially drawn here, to this board, because I was enchanted by the way a TV show had co-opted the internet to become an additional facet of the overall experience; all this enthusiasm, all this energy, all this creativity shared by anonymous folks drawn by a new approach to multi-media entertainment. . . . This maze stuff may turn out to be a crappy first experiment in how the internet can be integrated into a broadcast TV product. But the thing is it is sure to happen someday, and someday it will be done right, and we will find it hard to imagine how we were ever able to enjoy ourselves just watching TV. (NeillT006)

### Narrative Complexity and Media Convergence

*Lost* is, first and foremost, the television show that airs on ABC. Unfolding in linked episodes over several seasons, it offers an example of what Jason Mittell ("Narrative Complexity") calls a "new model of television storytelling . . . distinct for its use of narrative complexity as an alternative to the conventional and episodic serial forms that have typified most American television since its inception" (29). This new form of narrative complexity, Mittell argues, is characterized by oscillation between the demands of episodic and serial presentation; by self-conscious modes of narration that reflexively call attention to their own mechanics of plotting and production; and by narrative "spectacle," or complex narrative plot twists and turns (such as sudden revelations that make us question whatever has gone before—a narrative ploy frequently used in *Lost*). Such programs ask for new modes of audience engagement because of the complexity of both their content (plot, story world, characters) and their form and structure (for example, innovative use of flashbacks or self-reflexivity). Mittell points out how this narrative complexity is tied to "key transformations in media industries, technologies and audience behaviors" (30). VCR and DVD recorders allow viewers to control when they can watch (and re-watch) episodes. Internet sites allow fans to "embrace a 'collective intelligence' for information, interpretations, and discussions that invite participatory engagement." Other digital technologies, such as blogs, wikis, video games, and fan sites, enable viewers to extend their participation in these rich storyworlds" (31–32). But Mittell's point, like Jenkins's, is not so much the way technologies are converging as it is the kind of narrative competencies and forms of engagement these new converging technologies allow.

Unlike some of the television series Mittell discusses, *Lost* is made even more complex by a number of diegetic devices within the television broadcast that point viewers to other texts in other media that are also somehow within the *Lost* world. In particular, the network and producers have made extensive use of the Internet. An ABC podcast hosted by series writers and producers regularly appears, and there is an official forum, "The Fuselage" (www.thefuselage.com), for fan discussion. During the first season, a supposed diary by one of the survivors began to appear on the *Lost* page of the ABC Web site. Later in the first season, a tie-in Web site for Oceanic Airlines, containing "Easter eggs" and more information about the characters, came online. In the second season, more Web sites connected with the Dharma Initiative and other fictitious organizations and businesses mentioned on *Lost* began to emerge. And in the summer of 2006, ABC and Channel 4 (in the United Kingdom) produced The Lost Experience (TLE), an alternate reality game (ARG). In addition, three novels about *Lost* were published, including one by "Gary Troup," who was supposedly on Flight 815 and who later became a character in TLE.

As *Lost* converges on these various books, Web sites, podcasts, and forums, it becomes a form of what Jenkins (*Convergence*) calls transmedia storytelling,[2] a narrative that spills out into a range of multimodal texts that engage viewers in multiple ways. Partly this convergence is a way of branding *Lost* and generating more revenue through viral marketing. The *Lost*-related books, for example, are published by Hyperion, which is owned by ABC's parent company, Disney. Parts of TLE were sponsored and produced through affiliations with companies such as Verizon or Sprite. As one *Lost* fan is reported to have said, "You know you're at an official TLE site if you have to dig through a bunch of advertising to get to the good stuff" (Jensen). But media convergence in *Lost*, as in other transmedia narratives, is not just a question of technological proliferation. More important, perhaps, is that new ways to perform literacy are also converging.

### Narrative Desire: *Lost* on TV

At its best (and in my view *Lost* is an example of that best), transmedia storytelling takes advantage of the particular affordances of each medium that it uses. Thus the sustained hour of the television show is the primary conduit for developing what Peter Brooks calls narrative desire, a process of engagement in which readers and viewers are moved "forward, onward, through the text" (37). For Brooks, this act of narrative desire is achieved through the reader's response to the plot—that is, the ways in which readers are continually trying to make sense of the story's meanings as they progress through its textual representation. "Reading for the plot," in Brooks's terms, means discovering (or constructing) the intentions and connections among the discrete elements—incidents, episodes, actions—that make the story into a coherent whole. *Lost* is very good at evoking narrative desire in its fans, and this compulsion to know more is part of why they are so willing to follow its unfolding plot over various media. *Lost* exploits the narrative structure of many detective-type novels in which the story (the chronological sequence of events—what happened and in what order) is different from the discourse (the discursive sequence in which those events are represented). Understanding the extraordinarily complex plot of *Lost*—Steven Johnson ("Popular Culture") estimates that in its first year it concurrently engaged between thirty and forty mysteries—is the most significant activity undertaken by *Lost* viewers and the most important aspect of its popularity: it is the source from which all the other transmedia pleasures come.

The show is beautifully produced (its two-part pilot was one of the most expensive in network history), and its visual and aural qualities make it highly compelling. The landscape of the island is mysterious and evocative. The structure of each episode is complex—meaning viewers must watch carefully to establish connections among the show's various planes of action. In addition, viewers are confronted with multiple allusions and extratextual references. For example, there are frequent screen shots and mentions of literary works; several characters are named for philosophers (Locke, Rousseau, Hume, Burke, Bentham); and references to such discourses as religion, science, popular culture, and art history abound. Identifying these allusions (what viewers call a catch) and trying to connect them to the show's plot provide another significant interpretive challenge. Watching *Lost* on television—at least for most die-hard fans—is thus an intense, detail-oriented, cognitively challenging, and aesthetically engaging task.

In the television show, viewers are confronted with a densely realized dramatic achievement, a story that is deeply compelling on many levels. But there is another kind of pleasure in watching *Lost* on television that connects it not so much with dramatic genres (such as movies or plays) as with a different form of popular culture: games, particularly video games. Game allusions are sprinkled throughout the show. A bit of dialogue at the end of the fourth season, for example, sounds like it comes straight out of a game walk-through or the game Dungeons and Dragons.[3] Similarly, just as games frequently take the player from one "world" or "level" to another, each season of *Lost* opens up to a new world. The first season is set on the beach where the plane has crashed. The second explores an underground hatch built by the Dharma Initiative, and the third opens in the camp of the Others. The fourth shows the post-island world and closes with a character rotating a frozen wheel that, like similar devices in the game Myst, moves the action to a completely new level. Thus, along with its temporally complex plot and its discontinuities of story and discourse, *Lost* also exploits the kind of world-building spatial architecture that is characteristic of many video games (Jenkins, "Game Design"). The viewer, like the characters—and like a game player—explores spaces in the show, seeking clues to what is going on. Fans go through digital recordings of episodes, examining each frame of some sequences in order to uncover hidden details or decode particular signs. They then upload particularly significant screen shots to fan forums so that others can debate the scene's complexity. *Lost* producers seem aware of and to be aiming for this kind of viewing; at one point, after some of the survivors have discovered a film that supposedly explains the Dharma Initiative, one of the characters says, "we have to watch that again," a statement that many see as a nod to the repeated-viewing habits of die-hard fans. This experience of clicking through the show, frame by frame, is also reminiscent of playing early click-and-advance video games, such as Myst, in which one also goes screen by screen to try to orient herself in the game world, locate significant details, and map their relation.

*Lost* is also gamelike in the kinds of puzzles it poses. Like the characters who are trying to make sense of the strange experiences they have been thrust into, viewers try to solve the mysteries of the island and its mythology. To do this, they engage in a number of gamelike activities, such as identifying patterns, solving logical puzzles, and mapping mazelike spaces. In these and other ways, watching the television show itself seems to elicit in its viewers some of the same cognitive challenges (Gee; Johnson, *Everything*) and literate practices (Selfe and Hawisher) that characterize certain kinds of video games. For example, viewers have noted how frequently and in what ways they have seen juxtapositions of black and white, or how often they have observed objects that resemble an I-Ching symbol. They have spent enormous energy identifying every time one of the "numbers" appears on the island and in flashbacks (for example, "Sledgeweb's *Lost* Stuff") and trying to uncover the logic of their sequence. Fans have also drawn and debated maps of the Dharma hatches and the island's topography ("Diggs' Great LOST Map," for instance) or constructed elaborately detailed chronologies ("Lost/Timeline").

### Gaming Agency: *Lost* on the Internet

Gaming experiences are further enhanced for those viewers who supplement watching the television show with playing *Lost*-related games on the Internet. Here, viewers not only can find small tidbits of information, or "Easter eggs," that add to what they already know about plot or characters, but they also have the opportunity to experience the kind of gaming agency in which player actions have consequence within the game world.

These possibilities for gaming interactivity became even more exciting for fans in the summer of 2006 with the advent of TLE, a complex alternate reality game that was closely related to *Lost*'s unfolding plot and that played out over four months and over multiple media—including Web sites, podcasts, blogs, advertising, 1–800 phone numbers, press releases, and live appearances.[4] TLE is a good example of the collective intelligence and participatory culture that Jenkins (*Convergence*) sees as characteristic of transmedia narrative. The complications of the puzzles were simply too much for any one person, and it was only by pooling resources and sharing information in knowledge communities that groups of people were able to make their way through the webmaze of TLE. Therefore there were blogs, forums, and other Internet sites where people traded clues, outlined solutions, and competed to see who could crack the riddles first. TLE was not a complete success. For many viewers, including me, it became too complicated for insufficient payoff.[5] It also had to walk a delicate line. Because most viewers of *Lost* were not going to play TLE, nothing really crucial to their understanding of the show's plot could be uncovered. But, on the other hand, because some fans were investing enormous time and effort in playing the game, there had to be some reward. The result was that TLE solved a major puzzle for game players: The meaning of the numbers was revealed. But the solution to that puzzle, the numbers' significance, turned out to be irrelevant to the ongoing mysteries of the island as depicted in the television show—the version of *Lost* experienced by most of its millions of viewers. This compromise seems to have satisfied few, and much of the criticism leveled against TLE (for example, Jenkins) suggests some of the strategic and material difficulties that will need to be confronted as transmedia stories develop.

Although TLE was not a complete success, the gamelike strategies that are built into transmedia *Lost* do suggest that gaming is affecting how people learn to tell and respond to stories. Although there is a dense theoretical debate about whether games are narratives, a narrative/game like *Lost* may be a sign that such a debate is already outmoded.[6] That is, *Lost* (on TV and on the Internet) does not offer the choice between game and narrative; rather, the kind of narrative it tells requires gamelike responses to understand it. This kind of convergence goes beyond establishing connections among various forms of technological media. Instead, what may also be converging are the interpretive and performative commitments that belong to the genre of the video game and the genre of the television show. Media convergence can thus resemble the kinds of literate negotiations that are involved in other genre-blurring activities, such as interdisciplinary research or new media production. Furthermore, as more readers, viewers, and writers grow up on gaming, they may increasingly rely on gaming-type assumptions about how stories operate. If this is true, we can expect to see more narratives, like *Lost*, characterized by spatial structure, puzzle plots, interactivity, or other features associated with game play. Concomitantly transmedia productions such as *Lost* may help us understand how games can be made more effective and engaging through their narrative depth and density.

### Writing Back to *Lost*: Collective Intelligence and Participatory Culture

> "Four Quartets is a set of poems by T. S. Eliot. . . . The key repeating concept in the poem is 'The Still Point of the Turning World'—the idea of a central timeless concept or place of stillness around which all other things in the time revolve. I think the Island is an attempt by the

writers to literalize this idea into a physical setting. I believe there are intentional parallels and echoes between LOST and the quartets, which may help identify the show's central themes. I am going to deal with this one quartet at a time, due to length. I think each quartet represents a portion of the show (not necessarily chronologically)." (jmberger)

"Nice. I'm not familiar with the poem (or wasn't until now), but I've been thinking along similar lines. I was sort of thinking more of parallel universes, with the Island being a 'still point' that is removed from the infinite number of parallel universes, instead of from a single turning world." (zigbertToschius)

"Analyzing [this poem] is not unlike picking apart Lost. Your comparison is right on, brotha—but do you really, really think the writers were using this quartet as a blueprint for the show? I think eliot's universal ideas in the poems allow for the comparison—as we know the universal elements of Lost allow for many, many literary comparisons." (trinabobina)

Using the collective and participatory environment of the Internet, fans "write back" to *Lost* in a number of ways. There are the blogs and other sites that support TLE; the wiki *Lostpedia* (www.lostpedia.com) that functions as the series' unofficial encyclopedia; sites where fans post their own dialogue transcriptions, maps, diagrams, and other catalogs of *Lost* information. In addition, some viewers write fan fiction in which they extend the lives of the characters; create fan art, such as banners, avatars, or pictures; and produce video parodies and homages that they post on sites such as YouTube. And in what may be the deepest, most intense, and longest-lived kind of response, an extraordinary number of fans participate in forums where they gather to consider and debate what *Lost* "means." These sites, in particular, offer literate spaces in which people write and read together, in order to make sense of their shared experience of watching *Lost*. Here, in scenes of collective intelligence and participatory response, viewers construct and defend elaborate theories based on close readings of textual details. They also identify and speculate on intertextual references, probe the psychology of particular characters, and offer aesthetic evaluations. Among the many ways of writing back are suggesting scientific explanations drawn from physics, astronomy, mathematics, or biology and placing events and references into larger historical contexts. All this is done within a dialogic environment characterized by verbal play and intense audience awareness.

This is not to say that everything posted on fan forums is critically inflected—or even interesting. Much of what appears is community building:

trading quips, practicing one-upmanship, chastising those who egregiously break the rules. A great deal of what is offered as research or speculation is simply cut-and-pasting long sections from wikis or other sites. But there is also a high level of critical and intellectual engagement: a willingness to approach *Lost* as an intricate, multi-layered work and to understand its meaning and structure. In the sections that follow, I outline a specific set of critical activities I observed on one fan forum. My focus is on certain exchanges in which viewers talk about how *Lost* works as a complex narrative text. In particular, I identify three types of critical response: (1) close reading, in which participants work together to discover the formal and thematic coherence of specific scenes; (2) intertextual analysis, in which they collaborate to identify specific allusions and speculate on their significance; and (3) consideration of authorial intention, in which participants debate questions about interpretive freedom and textual authority. There are, of course, many other ways of interpreting *Lost*; posted responses also examine the show through science, history, politics, popular culture, philosophy, theology, art history, aesthetics, and from multiple other perspectives. In fact, many of the conversations on *Lost* forums resemble the kind of critical literacy acts prized in university classrooms. I chose to focus on literary analysis, though, because as an English teacher I am most interested in the ways people read and make sense of imaginative texts.

The exchanges I will describe all come from the Lost-TV forums (*www.losttv-forum.com*), an Internet site on which I have been an observer and occasional participant for about three and a half years. Lost-TV has been active since just after the show first began airing in 2005. As of December 2008 it had about 24,000 registered members, more than 2.5 million posts, more than 38,000 threads, and an incalculable number of hits.[7] Although it includes sections devoted to fan fiction, fan art, spoilers, TLE, the "webmaze," and various non-show topics, the exchanges I will consider all come from the General Discussion and Lost Theories sections, which represent the substantive core of *Lost* analysis and are the major sites for the critical activities I will describe.

### What Does This Mean? *Lost* and Close Reading

Reading the Lost-TV forums, one is struck by how carefully participants attend to *Lost*'s verbal, visual, and aural details. Their responses suggest an implied interpretive contract between them and the writers and producers that anything is potentially meaningful. Thus they comb through particular scenes, seeking details that might connect to or extend what they already know. Transcriptions of each episode are posted online, as are screen shots of key moments, allowing participants to do fine-grained analysis and highly recursive readings. Fans have noted inconsistencies that turned out to be prop errors, such as when a resume written in Korean appeared that suggested a character had begun working at too young an age. When this happens, fans are often irate because they believe the writers and producers of *Lost* have failed to live up to their end of the bargain. Similarly, although there is a great deal of theory talk on this forum (it is clear that some academics participate[8]), most fans do not accept the poststructuralist premise that the meaning of a text is indeterminate. Instead, many of them read like New Critics. They practice the close-reading strategies of New Criticism, with its focus on nuance and detail, and they share its belief that the successful work of art is a perfectly wrought artifact, in which every aspect has meaning and coherence.

As an example of this kind of close reading, I want to summarize a thread that was active during the second season, titled "The significance of the buildings in London (Pink Floyd Album Cover)." This exchange focused on trying to understand the meaning of a single shot of a building in London that was part of one character's flashback. Earlier, in a podcast, the show's producers had alerted fans that this episode would contain an image of an iconic building and some signage that would turn out to be important. When the shot was identified, it was noted that the building was also pictured on a Pink Floyd album called *Animals*. To try to understand why this was significant, viewers scrutinized the image in great detail and shared comments on what might be crucial visual details. They also discussed the names of the album songs and their possible connection to *Lost*. They considered the relation of the album to George Orwell's *Animal Farm* and the potential associations of that novel with *Lost* and noted that one of the songs is considered to be a parody of the Twenty-third Psalm (a reference very important in *Lost* history). They even analyzed the lyrics of the songs on the album in terms of their possible relation to characters. Interspersed with these comments were running tangential conversations about the Kinks (a discussion that was eventually moved, to the relief of many, to another thread) and about whether or not the second Pink Floyd band should have been allowed to use that name.

As the exchange continued, better images of the screen shot became available, and eventually viewers located the sign on the building and agreed that its second word was *Construction*. This led to speculation that the building or the sign was connected to Michael—a character who had revealed in an earlier episode that he used to be a construction worker. Fans then revisited key scenes from past episodes in which Michael appeared, trying to determine whether his actions, speech, or demeanor should now be interpreted differently. At this point, visual and verbal references to *The Wizard of Oz* were also beginning to emerge on *Lost*, and there was subsequently some discussion about whether the balloons on the Pink Floyd cover might be another allusion, or might even foreshadow the survivors' next attempt to get off the island via hot air balloon. Eventually, in a later podcast, the producers explained that the banner had been meant to be readable to viewers but was inadvertently made too small. They then announced that the banner said "Widmore Construction." Because, at that time, the name Widmore had no known associations with *Lost*, fans tried to figure out whether the word had a hidden meaning. There were several attempts to find anagrams, a type of puzzle frequently used in *Lost*, and in what even the writer of the post seemed to understand was a stretch, one viewer remarked, "This probably means nothing . . . but during the picture shoot for the Pink Floyd cover, the inflatable pig was either released or became unattached from one of the stacks . . . it drifted toward an airport and finally came down in Kent. There is a Widmore road/street in Kent."

This exchange is not the longest or the most intense moment of critical analysis on Lost-TV forums, but it is typical.[9] As an example of close reading, it has some problems. There is a tendency to go off topic, and some interpretations have only dubious warrants. There is also an often-strained sense of how much deliberation can reasonably be attributed to every element of such a complex production as a sixty-minute television show. Nevertheless, the exchange shows great attention to detail, a strong understanding of how a metaphoric chain of references can accrue meaning, insightful analysis of character development, knowledge of how music and words intersect, and an ability to recognize verbal play. And while it is possible to do this kind of close reading by oneself, it is not nearly as much fun. Nor is it as productive. Much of the information on which this bit of analysis was based came from media other than the show itself, including the podcasts in which the producers offered supplementary information, the uploaded screen shots that were meticulously scrutinized, the Web-based research about Pink Floyd, and, most important, the online conversations themselves. Viewers searched these converging media to collect evidence and advance claims. Through the dialogic interchanges of the forum, fans were able not just to put forward ideas, but also to test them in a critical environment where fellow participants offered additional or qualifying evidence, proposed competing explanations, and questioned the warrants for specific claims. These participants also debated the degree of intentionality that one might reasonably infer from a particular scene. Therefore converging media and collaborative participation in the forum deepened the acts of close reading in which fans engaged.

### Where Does This Come From? *Lost* and Intertextuality

The "Official List of Literature (Book/Author), Movie, TV, Song & Art References" posted on Lost-TV forums identified, as of 2006, about fifty titles that are "concretely or explicitly seen or mentioned" on *Lost*. About forty references are listed as having "looser associations / possible implied references or jokes," and there is a link to another post that compiles the "full definitive list" of "Character-created nicknames that are allusions to books, movies, TV." The "Official List" also catalogs the titles of about fifty songs (but "for brevity's sake, only songs heard prominently during the episodes, not all the songs of the soundtrack") and six works of art. As of 2006, the "The Lost Theories Index" listed at least seventy-five threads that built theories, including ones titled "Literature, Cinematic, Pop Culture & Music Analysis & Allusions" and "Mythology." (These lists are no longer kept current, and the actual number of texts alluded to in *Lost*, now at the end of its fourth season, is considerably larger.)

Some of these allusions are obvious. The Dharma Orientation film, for example, was hidden behind a copy of *The Turn of the Screw*, and the camera lingered on the cover of the book for several seconds. Many viewers suggested that the book's narrative complexity and its ambiguity about what really happened might suggest how viewers were intended to respond to the film-within-the-show hidden behind the book. Other references are more subtle. One location on the island, for instance, is called Pala Ferry. One viewer noted that Aldous Huxley's book *The Island* is set on an island called Pala, and others, seeking a connection, commented that the book not only tells the story of a utopian society that, like *Lost*, "grapples with ethic[s] and philosophy as themes" but also "includes references to Taoism, seen on the show."

Identifying and understanding these intertextual references is a major interpretive task for *Lost* fans. Indeed, there is great competition to be the first person to "catch" an allusion, although more merit is awarded to those who actually can make sense of it and relate it meaningfully to the show's plot or themes. The best responses are, predictably, to texts with which a great many viewers are already familiar, such as allusions to popular culture (music, films, novels) or to canonical texts that many have read or are reading in school. When the text is unfamiliar to most of the participants, and interpreters have to rely on plot summary or cut-and-paste paraphrase from an online site, the analysis is often thin. (One of the major references in the second season, for example, was to Charles Dickens's *Our Mutual Friend*, a lengthy book that almost no one on the board had read and that few, if any, were prepared to tackle.)[10]

The allusive, intertextual web of *Lost*, like the ludic maze of TLE, is a particularly rich site for participatory and collective response. The complex play of associations stirs what Jenkins (*Convergence*), writing about a similar effect in the film *The Matrix*, calls "epistemophilia" (98). This desire to *know*, one of the greatest pleasures that *Lost* so deftly evokes and controls, is made even more intense by shared interactions between diverse knowledge communities, many of whom know something others do not. That is, *Lost* constantly suggests to viewers that more is going on than meets the eye and that part of its complicated meaning lies in a dense network of proliferating associations. But there are simply far too many associations for any one person to identify, let alone understand adequately (think trying to read *Ulysses* by yourself). Thus viewers work together collaboratively to elicit, evaluate, and organize all this information, just as they work together to solve the other puzzles *Lost* offers.

Although responding to *Lost*'s intertextuality does resemble the collective responses of puzzle solving, it is more complicated because of the open-ended ways in which textual associations can multiply. If close reading asks *Lost* fans to be New Critics, sorting through its intertextuality makes them Deconstructionists. In a thread about a reference to the Ambrose Bierce story "An Occurrence at Owl Creek Bridge," for example, the discussion included not only summary and speculation about parallels in plot and character, but also a consideration of how the story related to other allusions in that episode, such as a snippet of Glen Miller playing "Moonlight Serenade" picked up by a character on a shortwave radio. This, in turn, led to discussion of an episode of *The Twilight Zone*, in which an "old man picks up an old radio signal from 20 years prior—a disc jockey introducing Glenn Miller's 'Moonlight Serenade.'"

Framing this discussion were questions about how intentional these linked references were and how viewers should interpret them. Participants debated, for example, whether they represented a specific set of clues (Ambrose Bierce and Glen Miller, like the *Lost* survivors, famously "went missing"), a thematic pattern, a particular nod to die-hard fans (the Bierce story takes place at the moment the main character dies, and there has long been speculation that *Lost* takes place in "purgatory"), or simply a way "to get us all to think in one direction without any consequence." When the exchange ended, the participants had identified and analyzed an intertextual web of allusions that was larger, more complex, and more ambiguous than any one person would have understood before the exchange began. One poster remarked, "That's why I love this board. Someone always knows what you don't."

### Who Is the Author? The Hermeneutic Challenges of *Lost*

One of the primary challenges for viewers of transmedia *Lost* is deciding which texts are "canon," a term that in fan culture refers to content considered genuine or official. The complications of determining what is canon are both logistical and hermeneutic. In particular, as "official" *Lost* spreads out across multiple media, especially the Internet, numerous fake sites have sprung up, forcing viewers to discriminate between counterfeit and legitimate information. But questions of determining authenticity go beyond figuring out the source code. Instead, fans engage in a process similar to what Jenkins (*Convergence*) describes as "exercises in popular epistemology" (44)—complex debates about who is the "author" of *Lost* and how viewers can discern authorial intent. Considering these questions is central to interpreting *Lost:* If it is true that any detail is potentially significant in revealing the mysteries, then it becomes imperative to determine which details, as canon, come with an author's imprimatur.

For some viewers, canon is anything originating with ABC—Web sites, blogs, podcasts, or interviews that originate from an ABC or Disney source. But others argue that a television network's corporate presence—especially its marketing strategies—is not necessarily under the control of the producers and writers of a particular show. These viewers require a specific endorsement by producers or writers before accepting information as canonical. But because a television series is the shared responsibility of many people—including producers, writers, directors, actors, musicians, cinematographers, and editors—many fans refer not to specific individuals but to a collective entity called The Powers That Be (TPTB). Identifying bona fide TPTB and discerning their intention is a significant challenge. In *Lost*, for example, fans raise questions about who writes the material on the complicated series of linking Internet sites of the webmaze, how much control TPTB have over the webmaze, and whether every link on an authorized Web site is also authorized. Fans also try to determine how much material is related to marketing and how much to advancing the narrative.

Such questions became particularly compelling at the end of the second season, when TLE appeared and when fans learned that the solution to certain questions—such as what the numbers mean—would be revealed there. The status of information gained from TLE thus became the subject of intense debate. Many of the most heated exchanges centered on whether participants could use information gained from TLE in their contributions to the subforums General Discussion or *Lost* Theories, or whether all TLE-related material had to remain in its own subforum, Lost Spin-off and Webmaze Discussion (see, for example, the fifty-plus-page discussions "I think we need to discuss . . . ," "mods we need a new forum," or "Lost and the Lost Experience inextricably entwined"). Many participants considered this a bureaucratic issue, but it turned out to have profound implications for how *Lost* was interpreted and how such critical concepts as "author" and "intentionality" were defined.

The participants divided loosely into two groups. One group, whose members might be considered purists, wanted only to watch the television show or accepted a Web page only if it bore the "signature of someone related to *Lost*." Because alternate reality games, by their very nature, do not announce themselves as games, however, they will always lack direct authorial attribution. The other group, which sometimes referred to its members as "mazers," wanted to use non-show material to build interpretations, although they differed in the standards they applied for establishing a definitive link to TPTB.

What ensued as these two groups hammered the issue out was a kind of theory war about the nature of authorial intention, validity, evidence, and interpretive freedom. Thus mazers, for example, pointed to the fact that several people known to be affiliated with the show had indicated that they were extending *Lost* onto Internet sites, while purists countered (arguing, one said, "from LitCrit 101") that what an author says he is doing and what he actually does are two different things. Moreover, since many fans are not certain that TPTB actually have a grand plan (as opposed to making it up as they go along), they question how much value to give TLE information, even if it can be proven to be canon. If, for example, TLE really did reveal the meaning of "the numbers," then this major piece of information ought somehow to be incorporated into the unfolding narrative of the television show—which thus far has not been the case. And if it is not included in the television show, then is the solution—and by extension the numbers themselves—in some sense now irrelevant to the main story? Fans thus become caught in a circle of interpretation. Or as one participant summed it up: "My theory is that this argument is metaphorical for the dissension we are seeing amongst the Losties themselves. On one hand we have the science based side—sorta Jack's side—the side that only wants to see what is shown to them and is therefore 'fact.' And then there are the 'faithful,' the Lockeans, the webmazers, who have seen what the island 'might' be and are enchanted by what the possibilities are" (lostagainnaturally).

At issue in the end was, in the words of another participant, whether "*Lost* is one product or two." If it is one, then the information offered on TLE is canon and potentially as relevant as anything on television. But if it is two, then TLE is a form of disinformation—no different from that offered by spoilers, previews, or even fan fiction. Clearly most of the millions of people who watch *Lost* did not play TLE, and for them the show and the ARG (which many viewers may not have been aware of) were distinct. But for fans of Lost-TV—including those who played TLE, those who only watched as others played it, and those who wanted to ignore the game altogether—the issues were more complicated. These discussions, Jenkins (*Convergence*) argues, "centering as much on how we know and how we evaluate what we know as on the information itself" (44), will become increasingly common as we learn to live in a knowledge culture. The debates about whether to use TLE information in General Discussion engage such epistemological issues, although in their concern with textual interpretation as well as factual information, they are even more complicated.

Implicit in these debates are even more complex epistemological issues about the nature of interpretive freedom. *Lost* generates responses—theories about what it all means—that are so complicated and so dependent on esoteric information that they could not possibly be the explanation for the action of a mainstream television show. At some level, fans must know that the mysteries of *Lost* will not, in the end, be tied up primarily by complex scientific phenomena such as Messier objects or abstract philosophical theories such as psychoanalysis or dialectical materialism. Television shows (especially those with large audiences) simply do not work that way. Nevertheless, many fans have claimed the kind of interpretive freedom readers have long asserted with literary texts such as novels or poems. The question of authorial intention is almost suspended as they construct ever more complicated frameworks in which *Lost* might be understood—simply, it would seem, for the sheer pleasure of doing so. This kind of participatory relation between the viewer and TPTB suggests that fans are willing to claim a share in interpretive ownership of the text—a move, I believe, that is enhanced in the participatory environment of convergence culture.

## Conclusion

Analysis of the literate acts involved in "viewing," "reading," and "writing back to" *Lost* supports the claims of Jenkins, Mittell, and others that more is converging in transmedia storytelling than the media themselves. Even more important are the new and dynamic interactions among "readers," "authors," and "texts." Readers converge with one another in their collaborative construction of meaning. Authors and readers converge as they interact on fan sites, and modes converge as meaning gets played out on visual and aural, as well as textual, planes. Texts converge as stories become instantiated on the page, on the Internet, and on television. These convergences are made possible by new media. But the change is not just technological; it is, as Jenkins emphasizes, cultural.

All this signals the beginning of a new, potentially rich, and still unpredictable kind of narrative. To address this innovative form of storytelling, we will need new theories of production and reception. In particular, we will need to pay attention to a set of complex interpretive and epistemological questions, such as: What constitutes a "text"? How do we define concepts such as "author" or "intention"? How do visual, aural, and textual modalities interact? What is the relation between low and high culture, as they are currently defined? But we should also consider what makes *Lost* so much fun. On Lost-TV forums, I witnessed many literate acts I would have welcomed in the classroom. I saw people not only devote enormous amounts of time and attention to understanding how a text works, but also respond to that text with intensity and passion. This is not an argument for replacing the study of literary texts with a study of a television show such as *Lost* (although I believe popular genres should be a part of our curricula). But it is an acknowledgment that many fans respond to *Lost* with the kind of commitment we hope to promote in university classrooms.

Why this deep level of engagement? I think *Lost* succeeds for many of the same reasons James Gee argues that video games are both personally compelling and highly effective in promoting learning. In particular, I am persuaded by Gee's "situated meaning principle," in which he claims that learning always happens in relation to embodied experience. I have argued elsewhere that one of the ways in which games make experiences feel embodied is through their narrative shape (Journet). That is, games narrate an imagined world in which the actions of the player—like those of the characters in the game—have consequence, and it is in relation to this story that learning becomes situated. The same is true, I believe, for the kind of learning promoted in *Lost*. It is in the context of *Lost*'s unfolding narrative that the interpretive actions of the viewer—like the physical actions of the characters—have meaning. What makes watching, reflecting on, or writing about *Lost* even richer than playing video games (at least for me), however, is that the challenges viewers take on are specifically literate.

These literate practices are both connected to and different from the ways we read and write about established narrative genres, such as novels. That is, responding to *Lost* requires much the same kind of interpretive work used with other types of imaginative texts, such as close reading, identifying intertextual references, and debating authorial intention. But it also calls on new literate challenges—such as moving among multiple media, discerning how narrative is shaped by media and mode, retrieving and sharing information and analyses in virtual environments, and building interpretations collaboratively in communities of participants. *Lost*, as an early attempt to construct a transmedia story, thus points to the new sorts of narrative experiences we may expect to find in convergence culture. As both a text to be read and a prompt for viewer and reader response, *Lost* suggests the kinds of interactive, collaborative, collective, and participatory literate practices that media convergence may promote.

## Notes

1. Ian Bogost's review of *Convergence Culture* offers a different argument about the role of narrative in media convergence. For a complex discussion among Jenkins, Bogost and others about these and further issues related to convergence culture in general, and *Lost* in particular, see "A Response to Ian Bogost," parts 1 and 2. For a more specific discussion about *Lost* as transmedia narrative, see Mittell ("Lost in an Alternate Reality;" "The Lost Experience").

2. Jenkins's main examples of transmedia storytelling in *Convergence Culture* are *The Matrix* movies and their associated short animated films, comics, and games. Other examples he discusses include *A.I.*, the promotional alternate reality game The Beast, and *The Blair Witch Project* with its Internet connections.

3. From a transcript posted in lostpedia.com: "You're gonna go into that greenhouse through that hole there. Once inside, you're gonna turn left. Go about 20 paces until you see a patch of anthuriums on your left. They're in an alcove against the north wall. Face the wall, reach down with your left hand. You'll find a switch that activates the elevator. The elevator takes you down to the actual Orchid station" ("There's No Place like Home, Part 1").

4. It is impossible to summarize fully the complications of The Lost Experience (TLE). The following synopsis draws on material from TLE posted on *Wikipedia*, where one can also find references to other sites that offer more detailed analysis of how the clues work. TLE began in May 2006, when advertisements on ABC led viewers to the Hanso Foundation's Web site, which contained clues to further Web activity. Parts of an interview with "Gary Troup" (the fictional author who had been on Oceanic Flight 815) then showed up on the Barnes and Noble and Amazon Web sites. Newspaper advertisements in real-world newspapers and television commercials subsequently directed viewers to further Web sites containing concealed messages about the author, his credentials, and his motives. After the second season finale of *Lost*, "Hugh McIntyre" of the Hanso Foundation appeared on *Jimmy Kimmel Live*. The same night, advertisements led viewers to other sites. In June a video was posted on Monster.com featuring a woman named Rachel Blake. Source code from the Hanso site then led viewers to a Web site run by radio host DJ Dan and to Rachel Blake's blog. In July,

Blake interrupted a Q&A at Comic-Con International that included *Lost* producers Damon Lindelof and Carlton Cuse to ask them about the Hanso Foundation. When Lindelof and Cuse said the Hanso Foundation was "fictional," Blake said that she had "evidence" that it really existed. Clues on Blake's blog and other Web sites then allowed viewers to start uncovering fragments of a video supposedly made by Blake that revealed nefarious activities of the Hanso Foundation. Eventually 70 codes or "glyphs" were released on Web sites and in physical locations, each corresponding to a video clip. The full video was then put on YouTube. In August, DJ Dan hosted a live webcast of his show to answer questions from actual viewers, as opposed to staged callers in earlier podcasts. Starting in August, free Apollo candy bars (a brand name first mentioned on the show) were handed out at other events. Viewers then got further messages from Blake, through email. TLE concluded with a "phone call" from Blake on the DJ Dan show and a link on ABC's home page to the full Rachel Blake video.

5. See Jensen for a good analysis of the ways in which *Lost* fans responded to TLE.

6. Henry Jenkins ("Game Design") describes the "blood feud" that has "threatened to erupt" between "the self-proclaimed ludologists, who want to see the focus shift onto the mechanics of game play, and the narratologists, who were interested in studying games alongside other storytelling media" (118). See also Nick Mountford or Marie-Laure Ryan for arguments that computer games are narrative and Espen Aarseth or Jesper Juul for arguments that they are not.

7. There have actually been more posts, but in 2005 Lost-TV was hacked and several months' worth of contributions were lost.

8. Most notably "drabauer," the forum name for Dr. Amy Bauer, an assistant professor of music at the University of California, Irvine, who is frequently interviewed about *Lost* and the forum and who edits *Lost Online Studies*, a peer-reviewed e-journal that is the organ for the Society for the Study of *Lost* (*www.loststudies.com*).

9. To get a sense of how complex and interesting discussion can be, see, for example, the threads "Freud Meets the Matrix," "Lost, time and cowboy movies (for Sergio Leone fans)," or "Official Numbers thread (aka How 'bout them numbers!!!)."

10. The relation of *Our Mutual Friend* to *Lost* turns out to be very complex. There are numerous plot parallels, and both works exploit the idea that people are often unknowingly connected to one another via different networks (the six degrees of separation principle). More important, though, the works are connected in their serial publication or performance and in the ways the authors have to help readers or viewers follow a complex plot over an extended period of time. Dickens's explanation of his aim and method at the end of *Our Mutual Friend* provides an apt description of the challenges of creating and responding to *Lost*: "To keep for a long time unsuspected, yet always working itself out, another purpose originating in that leading incident, and turning it to a pleasant and useful account at last, was at once the most interesting and the most difficult part of my design. Its difficulty was much enhanced by the mode of publication; for, it would be very unreasonable to expect that many readers, pursuing a story in portions from month to month through nineteen months will, until they have it before them complete, perceive the relations of its finer threads to the whole pattern which is always before the eyes of the story-weaver at his loom" (798).

## Works Cited

Aarseth, Espen. *Cybertext: Perspectives on Ergonomic Literature*. Baltimore: Johns Hopkins University Press, 1997.

"An Occurrence at Owl Creek Bridge." Online posting. 28 February 2006. Losttv-forum. Accessed 22 July 2008, http://www.losttv-forum.com/forum/showthread.php?t=14358.

Bogost, Ian. "Review of *Convergence Culture* by Henry Jenkins." *Water Cooler Games*, 1 August 2006. Accessed 22 July 2008, http://www.watercoolergames.org/archives/000590.shtml.

Brooks, Peter. *Reading for the Plot: Design and Intention in Narrative*. New York: Knopf, 1984.

Dickens, Charles. *Our Mutual Friend*. London: Penguin Books, 1997.

"Freud Meets the Matrix." Online posting. 7 July 2008. Losttv-forum. Accessed 22 July 2008, http://www.losttv-forum.com/forum/showthread.php?t=7434.

Gee, James Paul. *What Video Games Have to Teach Us about Learning and Literacy*. New York: Palgrave/Macmillan, 2003.

"I Think We Need to Discuss . . ." Online posting. 4 May 2006. Losttv-forum. Accessed 13 August 2009, http://web.mit.edu/comm-forum/forums/popular_culture.htm.

Jenkins, Henry. *Convergence Culture: Where Old and New Media Collide*. New York: New York University Press, 2006.

———. "Game Design as Narrative Architecture." In *First Person: New Media as Story, Performance, and Game*, edited by Noah Wardrip-Fruin and Pat Harrigan, 118–30. Cambridge, Mass.: MIT Press, 2004.

———. "A Response to Ian Bogost (Part One)." Online posting. 25 August 2006. Confessions of an Aca-Fan: The Personal Weblog of Henry Jenkins. Accessed 22 July 2008, http://www.henryjenkins.org/2006/08/hmm_buttery_a_response_to_ian.html#comments.

———. "A Response to Ian Bogost (Part Two)." Online posting. 25 August 2006. Confessions of an Aca-Fan: The Personal Weblog of Henry Jenkins. Accessed 22 July 2008, http://www.henryjenkins.org/2006/08/response_to_bogost_part_two.html.

Jensen, Doc. "'Lost' in the Video Blogs." *EW.com: Entertainment Weekly*, 2 July 2006. Accessed 22 July 2008, http://www.ew.com/ew/article/0,,1215341,00.html.

Jmberger. "The Still Point of a Turning World." Online Posting. 26 May 2006. Losttv-forum. Accessed 22 July 2008, http://www.losttv-forum.com/forum/showthread.php?t=30858.

Johnson, Steven. *Everything Bad Is Good for You: How Today's Culture Is Actually Making Us Smarter*. New York: Riverhead, 2005.

———. "Is Popular Culture Good for You?" *MIT Communications Forum*. Posted 6 October 2005, accessed 22 July 2008, http://web.mit.edu/comm-forum/forums/popular_culture.htm.

Journet, Debra. "Narrative, Action, and Learning: The Stories of *Myst*." In *Gaming Lives in the Twenty-First Century: Literate Connections*, edited by Cynthia L. Selfe and Gail E. Hawisher, 93–120. New York: Palgrave, 2007.

Juul, Jesper. "A Clash Between Game and Narrative." Posted April 2001, accessed 24 July 2008, http://www.jesperjuul.net/thesis/.
Kress, Gunther. *Literacy in the New Media Age.* New York: Routledge, 2003.
Lostagainnaturally. "Re: I Think We Need to Discuss . . ." Online posting. 4 May 2006. Losttv-forum. Accessed 22 July 2008, http://www.losttv-forum.com/forum/showthread.php?t=14328&page=18.
"Lost and the Lost Experience Inextricably Entwined." Online posting. 26 July 2006. Losttv-forum. Accessed 22 July 2008, http://www.losttv-forum.com/forum/showthread.php?t=22250.
"Lost Experience." *Wikipedia.* 6 July 2008. Accessed 22 July 2008, http://en.wikipedia.org/wiki/Lost_Experience.
"LOST Theories Index." Online posting. 27 October 2006. Losttv-forum. Accessed 22 July 2008, http://www.losttv-forum.com/forum/forumdisplay.php?f=97.
"Lost, Time and Cowboy Movies (for Sergio Leone Fans)." Online posting. 25 November 2007. Losttv-forum. Accessed 22 July 2008, http://www.losttv-forum.com/forum/showthread.php?t=24182.
Mittell, Jason. "The Lost Experience—Act II." Online posting. 20 July 2006. Convergence Culture Forum. Accessed 22 July 2008, http://www.convergenceculture.org/weblog/2006/07/the_lost_experience_act_ii.html.
———. "Lost in an Alternate Reality." *FLOWTV.* Posted 16 June 2006, accessed 22 July 2008, http://flowtv.org/?p=165.
———. "Narrative Complexity in Contemporary American Television." *The Velvet Light Trap* 58 (2006): 29–40.
"Mods We Need a New Forum." Online posting. 19 December 2005. Losttv-forum. Accessed 22 July 2008, http://www.losttv-forum.com/forum/showthread.php?t=11863.
Mountford, Nick. *Twisty Little Passages: An Approach to Interactive Fiction.* Cambridge, Mass.: MIT Press, 2003.
NeillT006. "Re: I think We Need to Discuss . . ." Online posting. 9 February 2006. Losttv-forum. Accessed 22 July 2008, http://www.losttv-forum.com/forum/showthread.php?t=14328.
"Official List of Literature (Book/Author), Movie, TV, Song & Art References." Online posting. 17 July 2008. Losttv-forum. Accessed 22 July 2008, http://www.losttv-forum.com/forum/showthread.php?t=16085.
"Official Numbers Thread (aka How 'bout Them Numbers!!!)." Online posting. 9 February 2008. Losttv-forum. Accessed 22 July 2008, http://www.losttv-forum.com/forum/showthread.php?t=7435.
"Re: What Did Eko See When He Opened the Bible?" Online posting. 1 December 2005. Losttv-forum. Accessed 22 July 2008, http://losttv-forum.com/forum/showthread.php?t=10437&page=2.
Ryan, Marie-Laure. "Beyond Myth and Metaphor: The Case for Narrative in Digital Media." *Game Studies* 1 (2001). Accessed 24 July 2008, http://www.gamestudies.org/0101/ryan/.
Selfe, Cynthia L., and Gail E. Hawisher, eds. *Gaming Lives in the Twenty-First Century: Literate Connections.* New York: Palgrave/Macmillan, 2007.

"Significance of the Buildings in London (Pink Floyd Album Cover)." Online posting. 24 March 2006. Losttv-forum. Accessed 22 July 2008, http://www.losttv-forum.com/forum/showthread.php?t=13707.

"There's No Place Like Home, Part 1." *Lostpedia*. Posted 16 May 2008, accessed 28 July 2008, http://www.lostpedia.com/wiki/There%27s_No_Place_Like_Home%2C_Part_1_transcript.

Trinabobina. "The Still Point of a Turning World." Online Posting. 26 May 2006. Losttv-forum. Accessed 22 July 2008, http://www.losttv-forum.com/forum/showthread.php?t=30858.

ZigbertToschius. "The Still Point of a Turning World." Online Posting. 26 May 2006. Losttv-forum. Accessed 22 July 2008, http://www.losttv-forum.com/forum/showthread.php?t=30858.

# Contributors

JOHN M. CARROLL is Edward Frymoyer Chair Professor of Information Sciences and Technology at the Pennsylvania State University. His research interests include methods and theory in human-computer interaction, particularly as applied to networking tools for collaborative learning and problem solving, and the design of interactive information systems. His books include *Foundations of Design in HCI* (2006), *Rationale-Based Software Engineering* (2008, with Burge, McCall, and Mistrik), and *Learning in Communities: Interdisciplinary Perspectives on Information Technology and Human Development* (2009). He serves on several editorial boards for journals, handbooks, and series and is editor in chief of the *ACM Transactions on Computer-Human Interactions*. He received the Joseph Rigo Award and the CHI Lifetime Achievement Award from ACM, the Silver Core Award from IFIP, and the Alfred N. Goldsmith Award from IEEE. He is a fellow of the ACM, IEEE, and the Human Factors and Ergonomics Society.

MARILYN COOPER is professor of humanities at Michigan Technological University, where she teaches graduate and undergraduate courses in writing, editing, grammar, and theories of writing, language, and pedagogy. She is currently working on a book entitled *The Animal Who Writes* in which she proposes that writing is a self-organizing system through which embodied beings interact with other sentient beings, material and semiotic resources, social and cultural structures, and biological and physical processes creating elaborate networks of living and that rhetorical agency and responsibility emerge from these networks. She is a past editor of *College Composition and Communication*. She co-authored *Writing as Social Action* (with Michael Holzman, 1989), and the Braddock Award–winning article "Moments of Argument: Agonistic Inquiry and Confrontational Cooperation" (with Dennis Lynch and Diana George, *CCC*, 1997).

PAUL HEILKER teaches courses in rhetoric, writing, and composition pedagogy at Virginia Tech, where he serves as codirector of the Ph.D. Program in Rhetoric and Writing and associate professor of English. He is the author of *The Essay: Theory and Pedagogy* (1996) and coeditor (with Peter Vandenberg) of *Keywords in Composition Studies* (1996), and his work has appeared in such venues as *College Composition and Communication*, *Rhetoric Review*, and *Computers and Composition*. He is currently at work on a critique of the spectrum metaphor that dominates discourses about autism.

JOHNDAN JOHNSON-EILOLA works in the Department of Communication and Media at Clarkson University, where he teaches courses in new media, design, and information architecture. He is the author of several books including *Datacloud: Toward a New Theory of Online Work* and *Writing New Media* (with Anne Frances Wysocki, Geoffrey Sirc, and Cynthia Selfe). He has also edited special journal issues on intellectual property (for *Computers and Composition*) and computer documentation (for the *Journal of Technical Writing and Communication*). He has won several awards for his teaching and research, including the 2005 Award for Best Collection of Essays in Technical or Scientific Communication from the National Council of Teachers of English and the 2005 Distinguished Book Award from *Computers and Composition*. His current research examines information flow in physical and virtual work spaces used by designers, video editors, and musicians.

DEBRA JOURNET is professor of English at the University of Louisville. One strand of her scholarship focuses on how narrative is used as a rhetorical and epistemic resource in the discourses of biology and composition. Her work in digital media explores how digital genres, such as video games, offer new forms of narrative engagement and open up new kinds of literate chapters. Her recent research on these topics appears in such journals as *Computers and Composition, Computers and Composition Online, Written Communication, Journal of Business and Technical Communication*, and *Social Epistemology* and in volumes from various presses. She is also the coeditor (with Beth Boehm and Cynthia Britt) of *Narrative Acts: Rhetoric, Race and Identity, Knowledge*, forthcoming.

M. JIMMIE KILLINGSWORTH is professor and head in the English Department at Texas A&M University and is the author or co-author of nine books and more than fifty scholarly articles and chapters. Killingsworth's teaching and research are focused on American literature, rhetoric, and cultural studies. In recent years he has concentrated particularly on eco-criticism and environmental rhetoric. His books include *Ecospeak: Rhetoric and Environmental Politics in America* (with Jacqueline Palmer, 1992), *Walt Whitman and the Earth: A Study in Ecopoetics* (2004), *Appeals in Modern Rhetoric: An Ordinary Language Approach* (2005), *The Cambridge Introduction to Walt Whitman* (2007), and *Reflections of the Brazos Valley* (with photographs by Gentry Steele, 2007). He is currently working on a general study of nature writing in modern times.

JASON KING is a Ph.D. candidate and Jim Corder Fellow at Texas Christian University, where he teaches developmental, first-year, and advanced composition courses. His research interests include rhetorical theory, literacy studies, disability studies, and new media writing pedagogy. His current research project examines the intersections between public rhetoric, literacy, and autism-parenthood.

CAROLYN R. MILLER is SAS Institute Distinguished Professor of Rhetoric and Technical Communication at North Carolina State University, where she has taught since 1973. She received her Ph.D. in Communication and Rhetoric from

Rensselaer Polytechnic Institute. Her research interests are in digital rhetoric, rhetorical theory, rhetorical genre studies, and the rhetoric of science and technology. Her publications have won three awards from the National Council of Teachers of English, as well as the Rigo Lifetime Achievement Award from the ACM-SIGDOC. Her university has designated her an outstanding teacher and an Alumni Distinguished Graduate Professor. She has lectured and taught in North America, Europe, and South America. She is a past president of the Rhetoric Society of America and current editor of *Rhetoric Society Quarterly*.

JAMES E. PORTER is a professor of English and Interactive Media Studies at Miami University, where he also directs the first-year composition program. Porter's research focuses on digital rhetoric, particularly on how the rhetorical topics of invention, delivery, audience, and ethics are changed in digital environments. He is completing a book, co-authored with Heidi McKee, titled *The Ethics of Internet Research: A Rhetorical, Case-Based Approach*.

STUART A. SELBER is an associate professor of English at Penn State, where he works in the graduate rhetoric program and also holds faculty positions in the College of Information Sciences and Technology and the Program in Science, Technology, and Society, which is jointly administered by the College of the Liberal Arts and the College of Engineering. His books include *Multiliteracies for a Digital Age* (2004) and *Central Works in Technical Communication* (2004). Selber has published work in a variety of journals, including *College Composition and Communication*, *College English*, *Computers and Composition*, *Technical Communication Quarterly*, and *Currents in Electronic Literacy*. He is currently working on an institutional study of academic computing.

GEOFFREY SIRC works in the English department at the University of Minnesota. His scholarly interest focuses on the histories and technologies of composition. His book, *English Composition as a Happening*, won the 2003 Ross Winterowd Award for outstanding book in the field.

SUSAN WELLS is a professor of English and director of first-year writing at Temple University in Philadelphia. She is the author of *The Dialectics of Representation* (1985), *Sweet Reason* (1996), *Out of the Dead House: Nineteenth-Century Women Physicians and the Writing of Medicine* (2001), and *Our Bodies, Ourselves and the Work of Writing* (2010). Wells received her doctorate from the University of Texas at Austin in 1978 and taught at the University of Louisville and at Wayne State University before coming to Temple.

ANNE FRANCES WYSOCKI is associate professor of English at the University of Wisconsin–Milwaukee, where she teaches undergraduate and graduate courses in written, visual, and digital rhetorics. She is lead author of *Writing New Media: Theory and Applications for Expanding the Teaching of Composition*, which won the Computers and Writing Distinguished Book Award. Her compositions have appeared in *Computers and Composition, Kairos, CCC,* and the *Journal of the Council of Writing Program Administrators,* as well as in many books. With Dennis Lynch she has published *Compose/Design/Advocate: A Rhetoric for Integrating Written, Visual, and Oral Communication* and *The DK Handbook.* She has designed and produced software to help undergraduates learn 3D visualization and to introduce them to geology. Her interactive new media pieces *A Bookling Monument* and *Leaved Life* have won, respectively, the Kairos Best Webtext award and the Institute for the Future of the Book's Born Digital Competition.

# Index

1960s, 51, 57, 69, 70, 138, 151–54, 156, 161, 169

Aarseth, Espen, 215n6
Abbey, Edward, 85–86
Abrams, M. H., 97, 99–100
accessibility standards, 188
activist scholarship, 131
activity theory, 17
Adorno, Theodor W., 29
Adriaansens, Alex, 108
aesthetics, 94–95, 97–106, 109–, 206; eighteenth-century notions of, 95, 97, 99
agonistic rhetoric, 124
Aigrain, Philippe, 176
Albrechtslund, Anders, 192n17
alternate reality games, 201, 203–4, 205, 206, 209, 211–12, 214–15n4
alternative press, 152–53, 156, 160
Amazon.com, 45, 47, 50, 177–78, 214n4
American Management Association, 52
Amerongen, Anton Van, 103
Anderson, Chris, 173, 176–77, 181
Anderson, Daniel, 193n21
Andre, Carl, 58, 59, 61–64, 68, 71, 72
animal communication, 21, 26
Aristotle, 113
Aronow, Zachary, 120
Ashbery, John, 60
Asperger's syndrome, 113, 114–15, 118
audience, 27, 68, 120, 121, 131, 159, 173, 174–75, 176, 178, 182–83, 185, 186, 189, 190, 200, 205

audiocassette, 64
authorial intent, 199, 206, 210–12, 213–14
autism, 113–31
autism communities, 116, 119, 124, 126–28, 130
autistic communities, 116, 119, 123–31

Bakhtin, Mikhail, 16
Baranowski, Matthew, 62
Barbrook, Richard, 186–87
Barlow, John Perry, 191n2
Barthes, Roland, 18
Basbanes, Nicholas A., 35
Bateson, Gregory, 16, 24, 26
Baumgarten, Alexander, 99
Bazerman, Charles, 17
Benjamin, Walter, 98, 105–6
Benkler, Yochai, 173, 178–80
Bergson, Henri, 16
Bettelheim, Bruno, 120, 124
Beyer, Catharine H., 62
bioregion, 84
Bir, Sara, 66, 67
Bitzer, Lloyd, 131
Blair, Carole, 94
Blockbuster (video/DVD store), 178
Bochner, Mel, 59, 70
bodies, rhetorical, 95
body, neglect of, 77, 88
Bogost, Ian, 214n1
books, cultural history of, 34
Boston Women's Health Book Collective, 164

## Index

Bourdieu, Pierre, 16, 99, 104, 176, 191n4
Bourdon, David, 59, 63
Braun, Colin, 104–5
Brooks, Peter, 201
Brouwer, Joke, 108
Brown, Greg, 88
Brown, John Seely, 189
Buell, Lawrence, 85
Burke, Kenneth, 114, 173

Capra, Fritjof, 29n1
Carr, Nicholas, 69–70
Carroll, John M., 135, 136, 140, 143
Carson, Anne, 67
Carson, Rachel, 82, 87, 91n5
Carter, Locke, 174
Cassirer, Ernst, 99
Centers for Disease Control and Prevention, 114
chaos theory, 16
Chen, Eric, 125
civil rights movement, 152, 158–59
Clark, Andy, 25–26
class, 89
Classen, Constance, 104
close reading, 206–8, 210, 214
cognitive ecologies, 18, 25, 28
collaborative construction of meaning, 213
collaborative environment, 143
collective intelligence, 198, 200, 203, 205
combinatory logic, 63, 67
community informatics, 143
complexity theory, 16
composition. *See* writing
composition, multimodal, 27
computer literacy, 28
computers and composition, 77
Computers and Writing Conference, 79
contextual inquiry, 183

convergence culture, 198, 208, 214
Cormode, Graham, 191n1
critical literacy, 90, 206
crowd wisdom, 173–74, 180–82, 184, 186, 187–88, 192n12
Crowley, Sharon, 78, 89, 90, 91n4
Csányi, Vilmos, 18
Cubbison, Laurie, 181
cultural commons, 180, 185
cultural evolution, 24
cyberpunk, 82
cyborg, 26, 77

Dartmouth Study of Student Writing, 60
database, 44–45, 47, 50–51, 180
Davies, Char, 96–97, 101
Davydov, Vassily, 17
De Vries, Hans, 103
Deconstruction, 210
Deleuze, Gilles, 16, 17, 33
Delicious Library, 40, 50–51, 54n4, 185, 192n11
delivery, 174–75, 181, 184, 190–91, 192n16
Derrida, Jacques, 25
design representation, 135, 136
DEVONthink, 18–19, 20, 22
DeVoss, Dànielle Nicole, 191n3
Dickens, Charles, 215n10
Dickinson, Greg, 94
différance, 25
digital art, 95–98, 101–3, 105, 107–8
digital economy, 173–77, 185, 188, 190
Dillard, Annie, 87
Diller, Elizabeth, 97
disability, 82, 113, 117
disability studies, 125
DiSessa, Andrea, 90
DJ Yoda, 67
Dobrin, Sid, 77
domain knowledge, 144

Domhoff, William, 159
Down's syndrome, 117
*Dr. Strangelove*, 138
Dreamweaver (software), 179–80, 191n7, 192n19
Duguid, Paul, 189
Dynabook, 134

Eagleton, Terry, 99–101
eBay, 178
eco-composition, 88
Economic Research and Action Project (ERAP), 158
economics, 173–75, 177–79, 185, 191
eco-rhetoric, 77–79, 81–85, 87
Edward Hoagland, 15
Elbow, Peter, 128, 130
Ellsworth, Elizabeth, 27
embodiment, 82, 84
emotional appeals, 95
Engeström, Yrjö, 29n2
environmental impacts of technology, 88
ethics, 53, 95, 102, 104–5, 188
ethos, 167
Eyman, Douglas, 175

Facebook, 48, 176, 192n17
Farooq, Umer, 146
feminism, 151, 156, 161, 164
fibromyalgia, 181
Flannery, Kathryn Thoms, 156
Flavin, Dan, 58
Flemming, David, 94
Flickr, 105, 110, 173, 185, 192n11
Flores, Fernando, 15
folksonomy, 180–82, 184–85, 189, 192n11
Foster, Hal, 71
Foucault, Michel, 33
Freud, Sigmund, 80–82
Friend, Tim, 23

Fritz, Gregory K., 119
Fumaroli, Marc, 114, 120

Gaonkar, Dilip, 113
Gaver, William W., 151
Gee, James Paul, 83, 89–90, 91n4, 203, 213
gender, 89–90, 126, 141, 151–52
genre, 27–28, 47, 57, 61–62, 65–68, 151–52, 155–57, 159–62, 165, 167–70, 202, 204, 213
Genung, John F., 56–57
Gibson, James, 151
Gibson, William, 82
Giddens, Anthony, 16
Gillmore, Gerald, 62
Gleick, James, 16
Godin, Seth, 45–46
Goff, Fred, 158–59
Golder, Scott A., 189
Goldhaber, Michael H., 174
Goodall, George, 34
Google, 40, 45, 47, 49, 51, 52–53, 70, 124, 178, 192n17
Gould, Stephen Jay, 24
Graafstra, Amal, 38
graffiti, 161
Grau, Oliver, 97–98, 102, 105, 107
Graves, Lucas, 161
Greenfield, Adam, 33, 38
Griffin, Donald, 18, 21
Guattari, Félix, 16, 17, 33

Halberstam, Judith, 91n4
Halloran, S. Michael, 94
Hanley, Daniel F., 105
Hansen, Mark, 16, 96–97, 102–3, 105, 107–9
Haraway, Donna, 77, 91n4
Harkin, Maureen, 99, 101
Harrington, Anne, 29n1
Hart-Davidson, William, 193n21
Hauser, Gerard A., 114

Hauser, Marc D., 18
Hawhee, Debra, 91n6, 94
Hawisher, Gail E., 89, 203
Hawk, Byron, 30n3
Hawley, Nancy Miriam, 167
Hayles, N. Katherine, 16, 29, 83, 91n4
Heffernan, Virginia, 70–71
Heidegger, Martin, 16, 19, 29, 79, 127
Hitachi's μ-Chip, 44
Hoffman, Keith M., 193n21
Horkheimer, Max, 29
Hornby, Nick, 66
Howes, David, 104
Huberman, Bernardo A., 189
Husserl, Edmund, 16
Hutchby, Ian, 151, 161
Hutchins, Edwin, 16, 26

IBM, 141
identity, 83, 85, 89, 94, 101, 144–45, 168–69
Ingold, Tim, 21–24, 29
institutions, 41, 98, 100, 164
Internet communities, 122
Internet textuality, 121
intertextuality, 205–6, 208–10, 214
invention, 20, 22, 24, 126, 175, 191
iPod, 134
Isenhour, Philip L., 143
iTunes, 67, 178

Jauneau, Roger, 153
Jay, Martin, 99–100
Jenkins, Henry, 198, 200–202, 204, 209–10, 212–13, 214nn1–2, 215n6
Jensen, Doc, 201, 215n5
Jeppeson, Marsha S., 94
Jerz, Dennis, 42
Jewitt, Carey, 151
Johns Hopkins Comparative Nonprofit Sector Project, 144
Johnson, Mark, 86
Johnson, Robert R., 189

Johnson, Steven, 18–20, 202, 203
Johnson-Eilola, Johndan, 30n4, 174, 188–89, 193n21
Journet, Debra, 84, 90, 213
Judd, Donald, 58–59, 70
Juul, Jesper, 215n6

Kahn, Herman, 137–40
kairos, 170
Kanner, Leo, 119
Kant, Immanuel, 99, 101
Kawahira, Kazumi, 105
Kay, Alan, 134
Keller, Joel, 68
Killingsworth, M. Jimmie, 79, 85, 86, 89, 91n2
Kingsley, Emily Perl, 117
Kitzhaber, Albert, 60–61, 63, 66
knowledge work, 41, 187, 190
Kornbluth, Jesse, 154
Kramer, Hilton, 58
Kress, Gunther, 151, 199
Krishnamurthy, Balachander, 191n1

labor, oppressed, 186
labor, politics of, 89
Lakoff, George, 86
Lanham, Richard, 174–75
*Larry King Live*, 120
Latour, Bruno, 16, 24–25
Lave, Jean, 16, 27
Lemke, Jay, 30n3
Leonard, Andrew, 68
Leroi-Gourhan, André, 22, 25
Lessig, Lawrence, 191n2
LeWitt, Sol, 58
Liss, Jennifer, 116
List, Regina, 144
Livingston, Ira, 91n4
Longinus, 64, 66–67, 71
*Lost* (television program), 198–214, 214n1, 214–15n4, 215nn5, 8–10
Louis, Morris, 60

Lowe, Charlie, 42
Lozano-Hemmer, Rafael, 107–8
Luft, Andreas R., 105

MacFarquhar, Larissa, 60
MacInnes, Fraser, 106
Malevich, Kasimir, 63
Mandelbrot, Benoît, 15
Mann, Steve, 80, 82, 91n3
Marinetti, Filippo Tommaso, 105–6
Marx, Karl, 104, 188
Maturana, Humberto, 15–16, 21–22, 24
Maurello, Ralph, 153–54
Mazlish, Bruce, 77
McCloskey, Deirdre, 174
McClure, Michael, 154
McCrimmon, James, 57
McLuhan, Marshall, 80
memory, 25, 42, 87, 135, 175
Merleau-Ponty, Maurice, 16, 21, 103
metadata, 45, 181
metaphor, 85–86, 161
Meyer, James, 58, 59, 61–63
Miettinen, Reijo, 29n2
Miller, Carolyn R., 151
minimalism, 58–59, 61
Mississippi Freedom Schools, 157–58
Mittell, Jason, 200–201, 213, 214n1
mix tape, 64–69
Moore, Thurston, 64–65, 67
Morphy, Erica, 54n5
Morris, Robert, 58–60, 63, 69, 72
Morroia, Fabrizio, 34–35
Mountford, Nick, 215n6
Mountford, Roxanne, 94
MP3, 68–70, 72, 134
MP3 blog, 69–70
multimodal texts, 199
Munster, Anna, 97, 102, 105, 107
MySpace, 176
Myst (virtual game), 84, 202–3

Nardi, Bonnie A., 191n5, 193n21
narrative desire, 201
National Institutes of Health, 114
National Science Foundation, 141–43
nature writing, 84, 87
Ne'eman, Ari, 117, 119, 125–26
Nelson, Amy, 123–24
Nelson, Theodor, 51
Netflix, 178
New Criticism, 207
New Left, 159, 161–62
New London Group, 16–17, 24, 27, 30n6
new media, 77, 80, 97–98, 102, 169, 170, 193n21, 204, 213
new media art. *See* digital art
New York University Child Study Center, 118
Nielsen, Jakob, 192n9
Nietzsche, Friedrich, 17
Nintendo Wii, 105–7, 110
North American Congress on Latin America (NACLA), 158–59, 169

O'Doherty, Brian, 71
offset printing, 151–53, 155–56, 159, 161, 165, 167
Olan, Susan, 153
Oliver, Martin, 151, 161
online privacy, 52
O'Reilly, Tim, 189, 191n1
Orr, David, 90
Owens, Derek, 88

Palmer, Jacueline S., 79, 91n5
Panganiban, Naomi, 62
participant journalism, 152
participatory culture, 198–99, 203, 205
pasteup, 153–54
Paul, James, 64
Paulding light, 19–20
Peck, Abe, 154
perception, 97, 102, 105

Perlstein, Daniel, 168
Perpetua, Matthew, 70
Petersen, Søren Mørk, 192n17
petroglyphs, 39
Pfaffenberger, Bryan, 151
place versus space, 83
Pollock, Jackson, 58, 60, 63, 72
Porter, James E., 175, 191n3, 192n16
Porter, Joshua, 180, 189
posthumanism, 78, 83, 91n4
postmodernism, 33–34, 78, 83
power structure report, 156, 170
power structure research, 151–52, 157–62, 164–65, 167–69
Prigogine, Ilya, 15
Prior, Paul, 17
Project Xanadu, 51
prosthetic technologies, 19, 79–82
public rhetorics, 119
public sphere, 113–14
Pucci, Enrico, 94
punk, 64–65

Raaf, Sabrina, 95, 110n1
race, 89, 126
Ramelli, Agostino, 34, 36, 54n1
Ramus, Peter, 175
Rappert, Brian, 151
Ratcliffe, Krista, 125–30
Ray, Janisse, 87, 91n5
Raymond, Eric S., 191n2
Reich, Robert, 189
Reid, Alex, 193n21
Reilly, Colleen A., 89
remediation, 45, 178, 190
reusable information objects, 178
RFIDs (radio frequency identification tags), 44, 49
Rhapsody, 177–78, 191
rhetoric of reconciliation, 125
rhetorical listening, 125–31
rhetorical situation, 26–27
Rice, Jeff, 193n21

Richards, I. A., 85
Ripeanu, Matei, 192n9
Robbins, Tom, 154
Rockwell, Geoffery, 34–35
Romanticism, 78, 85, 99, 100
Rosson, Mary Beth, 136
Royster, Jacqueline Jones, 126
Ruchelman, Leonard, 160
Russell, David, 17
Ryan, Marie-Laure, 215n6
Rzucidlo, Susan F., 117–18

Salamon, Lester M., 144
Salpeter, Eliahu, 160
Salvo, Mike, 174
Sante, Luc, 64
Savage, Michael, 120
Scarry, Elaine, 83
scenarios, 135, 137–38, 140–41, 145–46
Schechter, Daniel, 158, 160
Schiphorst, Thecla, 97
Scholz, Trebor, 192n17
Schryer, Catherine, 168
Schwarz, Heinrich, 193n21
Scofidio, Ricardo, 97
Seeley, Thomas, 181
Selber, Stuart A., 28–30, 90
Selfe, Cynthia L., 27–28, 89, 203
Selfe, Richard, 77
Selzer, Jack, 91n6
sensuous engagements, 94, 97, 105, 107, 109
sensuous training, 104
seriality, 59, 63, 69, 71
Sermon, Paul, 97
Serra, Richard, 63
Shannon-Weaver model of communication, 189, 193n20
Shapiro, Helen, 158–59
Shepherd, Dawn, 151
Shields, Rob, 89
Shirky, Clay, 176, 189

shuffle technology, 65
Shumway, David, 99, 104
Shusterman, Richard, 103–4
Silver, James, 168
Smalltalk (inspector tool), 136–37
social movements, 58, 77, 122, 151–56, 158–59, 165, 167–68, 170, 181, 187
social networks, 173–74, 176, 178–80, 182, 184–86, 190, 192nn9, 12
software design, 137
Sokolowki, S. Wojciech, 144
Sonic Youth, 64
Sony Walkman, 64
*South Park* (television program), 79
specifications, 135–37, 139–41, 146
Stein, Gertrude, 109
Stella, Frank, 63
Stengers, Isabelle, 15
Sterling, Bruce, 37–40, 42, 45, 47, 54n1
Stevens, Wallace, 60
Stewart, Susan, 104
Stiegler, Bernard, 103
Stoller, Peter, 104
Strahav Monastery library, 34–35
strategic planning, 137–39
Students for a Democratic Society (SDS), 158
Stuever, Hank, 67
Surowiecki, James, 173, 181, 186, 192n12
sustainable development, 144–45
symbolic-analytic work, 42, 48, 189
Syverson, Margaret, 30n3

Tapscott, Don, 191n2
task analysis, 136
Taylor, Mark C., 16, 29n1
teaching with technology, 26, 53
technical communication, 182, 188–89, 193n21
technological access, 185–86

technological affordance, 151–53, 155–57, 159, 161–62, 164–65, 167, 169–70, 201
technological design, 27, 61, 97, 123, 135–46, 151, 173–74, 176, 179–80, 182, 184–86, 188–90
technological eroticism, 84, 90
Technorati, 47, 49–51
techno-rhetoric, 77–84, 86–90
Terranova, Tiziana, 186–87
Terry Winograd, 15
text: as agent, 37; as artifact, 39, 40; as gizmo, 40, 42; as product, 40–41; as spime, 38–40, 44, 47–54, 54n4
Thierry, Lauren, 116–17
Tinderbox (software), 40, 43–44
transmedia narrative, 198–99, 201–4, 210, 213–14, 214nn1–2
Trimbur, John, 188
Tripper, Jack, 65–66
Truitt, Anne, 58
Tuan, Yi-Fu, 83
Turkle, Sherry, 84, 90

Uexküll, Jakob von, 18
underground newspaper, 152, 170n1
underground press, 152, 154
Unix, 135
user forum, 178–80, 188, 191n7, 192nn9, 19
user manual, 178
user-centered design, 189
user-generated content, 173–74, 180, 185, 187

van Uchelen, Rod, 153–55
Varela, Francisco, 15–16, 24
Vershbow, Ben, 53
video games, 37, 84, 89–90, 105–7, 199–200, 202–3, 213; and gender, 89; and race, 89; and violence, 89, 106–7
Viegener, Matias, 67

Vinyard, Chris, 68
virtual school, 141–42
visual rhetoric, 122–23
Vitanza, Victor, 17–18
Vroon, Piet, 103

Waddell, Craig, 91n5
Waddington, David, 152
Waldrop, M. Mitchell, 29n1
Web 2.0, 173, 176, 180, 185, 189–90, 191n1, 192nn11, 17, 193n21
Weblog, 45–47, 49, 52, 176, 191n5
Welch, Kathleen, 191
Wenger, Etienne, 16, 27
Whitman, Walt, 84–85
Whittaker, Steve, 193n21
wicked problems, 189
WIDE Research Center, 174, 182, 193n21

Wikipedia, 176, 191n1, 192n9, 214n4
Williams, Anthony D., 191n2
Williams, Joseph John, 89
Williams, Raymond, 99–101
Wilson, Carl, 65–69
Wittgenstein, Ludwig, 16, 20–22
Wood, Victor, 160
Wright, Suzanne, 120–21
writing: as complex system, 16; as conventional essay, 56–57; as embodied practice, 17–18; as interaction, 22; as social action, 15; as web, 16. *See also* minimalism
Wysocki, Anne Frances, 27–28

YouTube, 70, 116, 173, 176, 185, 192n11, 192n17, 205, 215n4

www.ingramcontent.com/pod-product-compliance
Lightning Source LLC
Chambersburg PA
CBHW021352300426
44114CB00012B/1193